WE KNOW IT
WHEN
WE SEE IT

WE KNOW IT WHEN WE SEE IT

What the Neurobiology of Vision
Tells Us About How We Think

Richard Masland

BASIC BOOKS
New York

Basic Books
Hachette Book Group
1290 Avenue of the Americas, New York, NY 10104
www.basicbooks.com

Printed in the United States of America
First Edition: March 2020
Published by Basic Books, an imprint of Perseus Books, LLC, a subsidiary of
Hachette Book Group, Inc. The Basic Books name and logo is a trademark of
the Hachette Book Group.

The Hachette Speakers Bureau provides a wide range of authors for speaking events.
To find out more, go to www.hachettespeakersbureau.com or call (866) 376-6591.

The publisher is not responsible for websites (or their content) that are not owned
by the publisher.

Print book interior design by Linda Mark.

Library of Congress Cataloging-in-Publication Data
Names: Masland, Richard H., author.
Title: We know it when we see it: what the neurobiology of vision tells us about
 how we think / Richard Masland.
Description: First edition. | New York: Basic Books, 2020. | Includes bibliographical
 references and index.
Identifiers: LCCN 2019030406 | ISBN 9781541618503 (hardcover) |
 ISBN 9781541618497 (ebook)
Subjects: LCSH: Vision. | Visual pathways.
Classification: LCC QP475 .M275 2020 | DDC 612.8/4—dc23
LC record available at https://lccn.loc.gov/2019030406

ISBNs: 978-1-5416-1850-3 (hardcover), 978-1-5416-1849-7 (ebook)

LSC-C

10 9 8 7 6 5 4 3 2 1

Contents

Introduction

THIS IS A BOOK ABOUT HOW WE SEE. THINKERS HAVE PONDERED vision for a long time, but most of their ideas were, by modern standards, naive: the eye *is*, in fact, something like a camera, but there is a whole lot more to vision than that. It may seem natural and simple that we can recognize the face of a friend—so much so that the ancients didn't even identify it as a problem—but there is actually nothing simple about it. To truly understand vision, you have to understand more than just how our eyes work. You also must understand how our brains make sense of the outside world.

Paradoxically, brains are pretty slow; neurons and their synapses work millions of times more slowly than modern computers. Yet they beat computers at many perceptual tasks. You are able to recognize your child among the crowd on the playground in milliseconds. How does your brain do it? How does it take a blunt stimulus—a patch of light, a vibration in the air, a change of pressure on the skin—and give it meaning? We have only glimpses of the ways, but what we have learned is fascinating.

I have been a neuroscientist since I was twenty-five—before the discipline of neuroscience officially existed—and I care as much about it now as I did then. I've watched our understanding evolve, and I've participated in the work myself. The basic narrative of this book is "how vision works"—from the retina to the highest visual centers deep in the temporal lobe. But I also want to let you follow the scientific journey, to see how basic neurobiology—not the talk-show kind—looks from beside the laboratory bench. So I'll mix in some scenes from the lab, and sketch some of the players.

We'll go through vision step by step. You'll hear that the world you see is not the world that actually exists: it has been broken into fragments by your retina and sent to your brain in separate channels, each telling the brain its specific little thing about the image. You'll learn how this recoding is accomplished by neurons in your retina, and why. We'll follow these signals into the brain, where they build our perceptions.

The brain holds many mysteries, but an important insight is that much of the brain works not by fixed point-to-point connections, like the telephone system, but by means of swarms of neurons interconnected, like a spiderweb, into nerve nets. These days, nerve nets are often associated with computers, but in fact they were thought up a half century ago by a far-seeing Canadian neuroscientist, Donald Hebb. A few years later the idea was co-opted by computer scientists. During the next decades nerve nets moved in and out of fashion, but better computers eventually allowed computer scientists to create the field of machine learning, better known as artificial intelligence. They showed that computer nerve nets can learn to perform dramatic feats, leading neuroscientists to look again at nerve nets in the brain. So today we have a remarkable alliance between neurobiology and computer science, each field informing the other.

Do brains use nerve nets to interpret the world? Does the brain work by "machine learning"? The answer seems to be yes—and brains do it a whole lot better than computers. To be sure, computers dazzle with certain of their feats—not just playing chess, but learning other, more complex tasks. Generally speaking, though, AI computers are one-trick ponies. And even the simplest require lots of hardware, with a concomitant need for lots of energy. In contrast, our little brains can do a multitude of tasks and use less energy than a nighttime reading light. Seen that way, computers are very bad brains, and a search is on to make them more brain-like.

The key to machine learning, as imagined long ago by Hebb, is that a nerve net connected by fixed wiring cannot do very much. Key is that the synapses that connect the neurons of a nerve net (or the simulated "neurons" of a computer) are modifiable by experience. This plasticity is a general rule in the brain—not just in sensory systems. It helps the brain recover from injury, and allows it to allocate extra brain resources to tasks that are particularly important. In vision, the nerve nets of the brain can learn to anticipate the identity of an object in the world—to supplement the raw information coming from the retina with its knowledge of images it has seen before. Boiled down, this means that much of perception is not just a fixed response to the visual scene but is learned. The brain's nerve nets recognize certain combinations of features when they see them.

Where does this lead in our search for understanding the actual experience of perception, thinking, emotion? We don't have a detailed answer, but we can see, far in the distance, how the final answer may look. Known, verifiable science can take us to an entry point. I will take us part of the way, to the seam where sensory experience turns into perception and thought.

Finally, where are "you" in all this? It's easy enough to talk about the brain as we see it from the outside, but where is the inner person that we imagine to be looking out through our eyes? There we can barely begin—and we run inexorably into the nature of consciousness, the self. We'll go there at the very end, with no answer but an attempt to see the problem more clearly.

PART I
THE FIRST STEPS TOWARD VISION

DURING THE 1960s, A GOOD TEACHER NAMED JACOB BECK GAVE A college course titled simply "Perception." The course met in a small auditorium tucked into a corner of Memorial Hall, a nineteenth-century brownstone colossus erected as a memorial to Harvard's Civil War dead. The lecture hall's gradual slope accommodated perhaps a hundred brown wooden desks, covered with a century's coats of yellowing varnish. A black chalkboard stretched the width of the front wall. High on the left wall were sparse windows. The room was otherwise lit by a few incandescent bulbs, turning the auditorium a soft yellow. Thirty or forty students were thinly scattered around the room.

Beck was as straightforward a teacher as the name of his course would suggest. His manner was pleasant enough, but he was not particularly interested in charming students—his main mission was to present his material in a clear and organized way. He used careful notes and stuck to them. He spent the first few minutes of each lecture reviewing the main points covered in the previous one.

Beck did not need showmanship. The material was fascinating in itself. To be sure, he taught us the basics: pressure on the skin deforms a nerve ending, which sends a signal up the spinal cord to the brain. Some of our skin sensors signal light touch, some signal heat, and some are for things moving across our skin—say, a venomous bug dropping on your arm from the forest canopy. Facts like these were interesting in their own right. But the most wondrous thing of all—the great challenge Beck posed to his roomful of nineteen-year-olds—was object recognition.

On one hand, this is a problem of sensation—how the eye works, how it signals to the brain. But it connects with the great issues of perception: thinking, memory, the nature of consciousness. We can get our hands on the pathways of sensation. We can record the electrical signals in sensory pathways. We can tease the neurons to tell us what they see. We now know a lot about how the sensory signals are handled—how they are passed from station to station in the brain. This gives us a handle on the larger questions; it is a place where we have certain kinds of secure knowledge. We are only starting to understand where the brain takes things from there. But taking vision step by step gives us a platform from which to peer toward the great mysteries.

1 | The Wonder of Perception

> The pears are not viols,
> Nudes or bottles.
> They resemble nothing else.
>
> They are yellow forms
> Composed of curves
> Bulging toward the base.
> They are touched red.
>
> —WALLACE STEVENS

CONSIDER THESE THREE FACES. ALTHOUGH THE IMAGES ARE slightly blurry and the contrast is poor, you can tell them apart. The woman on the right has a slightly rounder face; the boy on the left has a strong chin. If they were your son or your daughter, your friend or your mother, you would recognize them across

an amazing variety of situations. You would recognize them in plain clothes, without their makeup. You would recognize them in front of you or from an angle. You would recognize them in bright light or dim, nearby or at a distance, glad or sad, laughing or silent.

Yet *how* do you recognize them in all those different instances? The actual image that falls upon your retina is physically different in each case. Your brain adjusts to each version: larger or smaller, brighter or dimmer, smiling or glum. The permutations of faces, received as physical stimuli falling on your retina, are almost infinite. Yet you recognize familiar faces instantly, without effort. And you can tell apart not just these three but hundreds or thousands of faces. How can the brain—which is only a physical machine, like any other—perform this task so well?

It may help to think about a simpler example. Imagine that you must design a computer program that can recognize the letter A. Modern computers do this with ease, right? But in comparison with brains, they cheat. More on that in a minute.

The solution seems obvious: somewhere in the computer (or your brain) there must be a map or template of the letter A. Then the computer (or the brain) can just compare an A with the template and match them. But what if the size of the A to be recognized is different from the size of the template? The computer (or the brain) would have to conclude that they are not the same letter.

Well, why not just have the computer test a bunch of different-sized templates? That would fix the problem:

$$\text{A} \quad \text{A} \quad \text{A} \quad \overset{\checkmark}{\text{A}} \quad \text{A} \quad \text{A} \text{A}$$

No doubt about it, that would work. Suppose, however, that the test A is now tipped a bit: A + A → A They won't match, no matter how perfectly the computer has guessed the size.

OK, then, let's have the computer compare against all possible sizes and all possible angles. If the computer is fairly fast, that might work. But in the end, we'd have too many variants—line thickness, color, font, and so on. And then we'd have to multiply all those variants by each other. The computer ends up having to test all possible sizes *times* all possible angles *times* all possible fonts *times* all possible colors, and so on. The number of combinations that has to be tested becomes very, very large, impractically large. All this hassle for a simple letter.

With faces, there is almost no limit to the variants. A face can be smiling or frowning, dim or bright, viewed from the front or at an angle. And the components of brains—neurons and synapses—are, compared with computers, very slow. It takes about a thousandth of a second for a neuron in a human brain to transmit its most basic signal across a synapse to one of its fellow neurons. During that time, a pretty fast modern computer performs something like one million operations. This superhuman speed is why I said that computers cheat—they do something that ordinary wet biology could never do. Say it takes one hundred operations for a computer to make one comparison. A computer could therefore make a hundred thousand comparisons in the time it takes a brain to transmit a single nerve impulse across a synapse. And that's not counting the time it takes the signal to travel down the nerve fibers that connect neurons. If it were making comparisons in the same way computers do, your poor old brain would take minutes to recognize even the most familiar face. In other words, making lots of guesses is not an option for brains.

Here's another example, drawn from a different sense—hearing.[1] It's the problem called segmentation. If I say to you "The dog is blue," you'll generally hear the words as they are written on this page. But normal spoken speech does not have breaks between words. In the actual acoustics of that sentence (unless you speak with artificial breaks), there are no empty spaces between the sounds "the," "dog," "is," and "blue." In physical reality, the sentence is a single long sound.

To make sense of it, our brains break that long sound into words we know from a lifetime of speaking English (or whatever language we are using).

Once again, it is virtually impossible to see how the brain could use a template and match words against it. How many sounds would the template include? Certainly far more than the words in a dictionary. And this is to say nothing of different accents, rates of speaking, background noise, and more. So the brain isn't using a template to understand this string of sounds.

This whole big mystery—an act that we perform many times daily with such ease—is termed the problem of object recognition. We think of it as being about sensory experience, but it is just as much a problem of memory: object recognition is matching a present stimulus to the memory of an object seen in the past. To figure out how it works is a spectacular technical challenge—the Mount Everest of sensory neurobiology.

2 | Neurons That Sing to the Brain

> You learn something general by studying
> something specific.
> — STEPHEN KUFFLER

I HAVE TOLD YOU THAT THE WORLD YOU THINK YOU SEE IS NOT THE world that actually exists. It has been altered by your retina, fragmented into dozens of different signals for transmission to the brain. The retina parses the visual image into its most telling components and sends a separate stream of signals about each of them to the brain. The rest is ignored, treated as background noise. This kind of stripped-down signaling, which is a search for economy you'll hear more about, is not just a result of evolution amusing itself; it is one of the most fundamental principles of all perception.

To see how it happens, we have to get down to basics.

A SINGLE NEURON

A neuron is not a complicated thing. It is a physical object, albeit a very small one, made of materials we understand. It has the normal parts that make up any animal cell, with only a few unique features.

When you concatenate a few hundred million neurons, though, big things happen: recognizing a friend, hearing Beethoven, a one-handed catch of a thirty-yard forward pass.

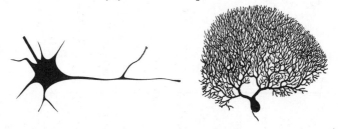

A neuron, like all vertebrate cells, is a bag of water separated from the surrounding water by a thin, fluid membrane. The membrane serves to divide the space inside the cell (black in these drawings) from everything outside it. A few neurons are more or less round, like a kid's balloon. Others take more complex, amoeboid shapes. Still others can have bizarre and complicated arrangements. Many neurons look like skeletons—a tree in winter. The twigs and branches reflect the neuron's connections with its near or far neighbors. No matter how baroque the shape, however, the cell always consists of a single space enclosed by a membrane. Its thin twigs enclose long thin spaces, like branching, convoluted soda straws.

What is this cell membrane? It is made of lipid, a variety of fat. Since fat and water do not mix, the cell membrane stays separate, a bit like a soap bubble. By itself, the cell membrane cannot accomplish much of anything. In the lab you can make an artificial cell that has only a cell membrane. Such a cell just sits there. An actual cell membrane is studded with a myriad of fancy little machines that do specific tasks—for example, embedded protein molecules that sense other molecules impinging from outside and then open a gate between the outside and the inside of the cell, which allows an electrical charge to pass. This is the basis of the nerve impulse.

While nerves have an impressive functional repertoire, their main function—the function that distinguishes them from almost all other cells—is to communicate with other neurons. In most cases they do so by transmitting brief impulses of electrical activity, known as spikes. Spikes can travel short distances or long. Some neurons talk to others (we say they "conduct nerve impulses") only within their own restricted neighborhood. These so-called interneurons (local circuit neurons) signal over distances as small as 10 micrometers, which is just a hundredth of a millimeter. Alternatively, some spikes travel all the way from your brain to the bottom of your spinal cord, as when you seek to wiggle your big toe, or in the reverse direction, as when you stub it on a brick.

Spikes are not electrical currents, like the currents that a copper wire conducts. They are a more complicated biological event, in which the cell membrane actively participates: they are the electrical reflection of the movements of charged ions into and out of the cell, guided by specialized proteins that sit in the cell membrane. For that reason, they travel very slowly compared to electrical conduction down a wire. Nerve impulses travel along an axon at a speed ranging from roughly 10 to 100 meters per second, depending on the axon. Electricity flows down a wire at about 300 million meters per second. From the point of view of our brain's ability to compute things, this slowness of conduction is a big deal. It is the main reason brains cannot use brute-force, dumb strategies to solve problems.

At the end of the axon is usually a synapse. A synapse allows one neuron to talk to other neurons across the gap that separates them. At the synapse, an electrical signal in one neuron is changed into a chemical signal; specialized synaptic machinery allows the spike to trigger the release of chemicals that are sensed by the second neuron. These are neurotransmitters, about which we hear so much in the news. Because there are lots of different kinds of neurotransmitters, used for different purposes at various places around the brain, and lots of steps involved in their release, this is a point where we can manipulate brain function—for therapeutic goals or for recreation.[1]

Nicotine acts on synapses. So do antipsychotic drugs, and those that control epileptic seizures. So does Valium, to make you calm, or Prozac, to make you happy.

A neurotransmitter released by one neuron can make another neuron more excited or less excited. (In reality, a neuron is rarely receiving just a single signal, but for the present purposes let's just assume it does.) The second neuron integrates all the inputs it receives. When enough impulses reach that neuron within a short time, what we call an "action potential" is triggered in the neuron. That action potential can propagate autonomously within the second neuron and excite or inhibit a third neuron, and so on.

At this point, we see the second big thing neurons do: they decide which inputs to pass on to further neurons and which inputs not to pass on. They make this decision solely by adding up all the inputs the cell receives. This is a bit of a simplification, as the ways in which inputs can be received are wonderfully varied. But to take a simple example, they add excitatory inputs and subtract inhibitory ones. The study of this process constitutes a field of its own within neurobiology; some of my smartest colleagues have spent their lives unraveling the many and elegant ways in which synaptic communication can occur.

Now, though, we'll think of neurons in their simplest mode, waiting for inputs to come along and firing an action potential when those inputs reach a certain magnitude. But just sending messages from neuron to neuron does not make a brain a brain. It is the combination of neuronal signaling and neuronal decision-making that makes a brain a brain. I'm simplifying because my task here is to tell you about perception. For that, we only need to understand a few things. The most important is that an action potential causes an electrical change wherever it goes. Critically for our story, that electrical change—the spike—can be eavesdropped upon by mortals armed with long thin probes called microelectrodes.

HOW SENSORY NEURONS SIGNAL

As I've said, neurons carry messages from place to place over a distance that can be short or long. In a giraffe, the neurons that control walking can span 2.5 meters, reaching from the brain to the lower spinal cord. In all but a few cases, however, the means of signaling is the same: somewhere on the surface of the cell there is a stimulus that initiates an action potential that spreads throughout the neuron.

All neurons that sense the outside world—whether through touch, hearing, vision, or smell—do the same fundamental thing: they detect an event in the world and transmit a signal about it, sometimes with a relay or two, to the brain. But they do this in quite different ways, mirroring the events in the world they are sensing, which are also physically different.

Consider the sense of touch. A perception of touch originates when the skin is deformed through pressure. This could occur by a finger stroking your wrist, a mosquito walking gingerly in search of a soft spot to stab, or your brusque collision with some solid object. These deforming pressures, hard or soft, are detected by nerve endings located just below the surface of the skin. Each ending is part of a neuron.

Skin	Spinal Cord	Brain

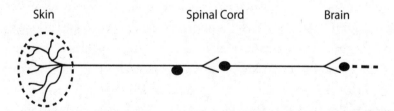

This image shows two neurons on the touch pathway; the patch of skin, shown by the dashed circle, is known as the receptive field. Information travels from left to right in this diagram. The first neuron has a long fiber (axon) that runs from a place on the skin—where the nerve ending forms many small branchlets—to the spinal cord. When, say, a mosquito lands on your arm, the mosquito's foot ever so slightly depresses the skin over the nerve ending. That pressure is transmitted to the neuron and a nerve impulse is initiated. That impulse travels along the axon, through the cell body and ends at a synapse (indicated by a forked line) upon another neuron, located in the spinal cord, which then projects to the brain. (Other pathways to the brain exist. This is just one of the simplest.)

The tactile neuron's branches can detect the indentation of the surface of the skin through what is known as a mechanosensitive ion channel. This channel is a protein in the cell membrane. Deforming the mechanosensitive channel allows positive ions to flow from outside the cell into the nerve ending. The flow of positive ions tends to excite the ending. When the excitation reaches a certain threshold, the ending begins to fire action potentials. These travel up the skin sensory nerve (axon) and past the cell body to a collecting site in the spinal cord, where the axon encounters a second neuron that will transmit the information toward the brain for interpretation. Note that the information from this skin sensory nerve has told the rest of the nervous system three things: that there is something touching your skin, that it is located just above your right wrist, and that the thing is fairly light.

First, the "where," which is easy. The endings of an individual tactile neuron in the skin cover a limited space on the skin. This space may be tiny, such as on the hand or lip, or broader, such as on the skin of the back. The brain knows what region each nerve surveys, and from that it knows where on the skin the stimulus fell—where the receptive field of that neuron is located.[2] Obviously, if the stimulus falls on an area of the body like your fingertip, which is covered by many tiny nerve endings, the brain will know more precisely where a small stimulus fell than for places with only a few huge endings, like on your back.

I introduced an important piece of nomenclature when I referred to the dashed circle in the diagram. I called this area under the terminal branchlets of a sensory axon a cell's receptive field. The receptive field is the specific part of the skin from which a particular sensory axon can be excited. As you will see, we use the same term to talk about vision, where "receptive field" refers to the area of the retina that excites a particular visual neuron—in the retina or later in the visual system.

Now for the "how much" question: how light or heavy the stimulus. How does the skin sensory nerve convey that? All sensory axons—for touch, hearing, vision, smell—communicate with the brain by a coded frequency of action potentials. A light touch produces only a few action potentials; a stronger one produces a more rapid string of them. That's how the brain, or an experimenter who can monitor the rate of firing, can tell how strong the stimulus was.

Many scientists (including me) have speculated in print that additional information may be contained in the detailed *pattern* of the action potentials, just as a pattern of key taps conveys information in Morse code.[3] The pattern could tell the brain, for example, what type of receptor a particular axon is carrying signals from (see the next paragraph). Certainly the pattern of spikes influences how the brain responds; we know that closely spaced action potentials (spikes) excite the postsynaptic cell more powerfully than widely spaced ones. But nobody has proposed and tested a specific code that has turned out to be convincing.

Even more interesting is the "what" part of our question. The brain wants to know: "What kind of a thing is touching my wrist?" All touches are not created equal. There are several different kinds of touch neurons, responding to different aspects of touch. One type of touch receptor is moderately sensitive to light touch on the surface of the skin and keeps sending a signal to the brain as long as the lightly touching thing is still touching. Another type of receptor responds only to fairly strong pressure and responds only at changes in touch—when the pressure first starts, or when it ends. Presently we know of more than a dozen kinds of primary touch neurons. These can be separately tested in a neurologist's office. That, in fact, is what she is doing when she compares your sensitivity to a pinprick with your sensitivity to a touch from a buzzing tuning fork.

Interestingly, many of the differences among skin receptors are due not to fundamental differences in the neurons but to the different structures in which the nerve endings are embedded. The endings of individual touch neurons are surrounded by specialized cell structures, and these cause the sensory neuron to respond to different kinds of touch. Think of the difference between a snare drumstick and a bass drumstick. Both are basically sticks, but one has at its business end only a small wooden ball, and the other has a great fuzzy pad. They correspondingly make different sounds when they hit the stretched skin of a drum. In addition, different receptors express different ion channels, giving their responses new richness. The details of these arrangements, though wonderful tributes to the skill of evolution, are not really important here. What matters is just that the different types of neurons respond to different aspects of the world's impacts upon the body: some respond to the jump of a flea, while others require the blow of a fist. Of course, there are many intermediate cases; most situations are signaled to the brain by a mixture of neuronal types. As one of the experts has written, "Like individual instruments in an orchestra, each [touch neuron] subtype conveys a specific feature of the forces acting on the skin, collectively culminating in a musical symphony of neural impulses that the brain translates as a touch."[4]

And this is a general principle for all sensory systems. Taste is subserved by a set of five types of taste buds, corresponding to sweet, sour, salty, bitter, and umami (a complex taste triggered by some amino acids). The sense of smell, remarkably, has by present count around four hundred types of receptor, each selectively sensitive to a particular volatile molecule. This accounts for the abilities of wine tasters to distinguish hundreds of wines by their bouquet (sadly, I lack this) and for the way a specific perfume can call up thoughts of an old lover.

HOW VISION IS LIKE TOUCH

I've gone into this much detail about how touch works because the basic principles of touch and vision are similar. All neurons work in essentially the same way. Both vision and touch boil down to the brain locating inputs that land on a sheet of sensory cells—the skin or the retina—and both involve a great variety of different sensors. In both cases, single neurons are tasked with telling the brain a few very specific things, and both touch and vision neurons respond to only a limited receptive field. But when it comes to vision, we know much more about how the receiving structures of the brain deal with input. Thus we can understand much more about how the brain interprets the symphony of signals arriving from the retina.

We have just seen that individual neurons innervating the skin tell the brain different things about objects that have touched it. The same fundamental principle underlies the workings of vision: each fiber of the optic nerve reports to the brain about one small region and one specific feature of the scene in front of you.

The retina is a microprocessor, like the one contained in your cell phone, your camera, or your wristwatch. It contains many different types of neurons, upon which we will lavish much attention later. For now, we'll just think about the retina's output, as conveyed by the activity of its long-distance signaling neurons, called retinal ganglion cells (analogous to the touch neurons that project up the spinal cord). Each human retina contains about one million retinal ganglion cells. These cells collect the inputs of several types of internal retinal neurons and run them up to the brain. Bundled together, the long axons of the retinal ganglion cells constitute the optic nerve.

The first in-depth study of retinal ganglion cells was done by the Hungarian American scientist Stephen Kuffler. Steve's long-term interest was elsewhere—the mechanism of synaptic transmission—but after migrating around the world during World War II, he landed a position in the Department of Ophthalmology at Johns Hopkins University. In part out of gratitude to his hosts, he did a study that to this day remains fundamental to our understanding of vision.

In about 1950 Kuffler recorded the electrical activity of single retinal ganglion cells in the eyes of deeply anesthetized cats. Having put a microelectrode into an eye, once he encountered a retinal ganglion cell he was able to study the train of impulses formed by that ganglion cell when he stimulated the surface of the retina with small spots of light. The spots of light needed to be small because he needed them to be like images cast on the retina by objects in the world. By the time they arrive at the retina, images are hugely demagnified—for example, the image of my thumbnail at arm's length spans only four-tenths of a millimeter on my retina.

The signals sent by the retinal ganglion cells Kuffler observed were a lot like the touch signals sent by skin sensory neurons. Each retinal ganglion cell is responsible for a single small patch of the retinal surface—its receptive field. The smallest of these areas in the cat's eye was around 40 micrometers, or 4/100 of a millimeter. While we don't know the size of individual receptive fields in humans—we have no medical reason to record from retinal ganglion cells in a person—indirect evidence indicates that our smallest receptive fields are about 10 micrometers in diameter. A Nobel Prize winner has calculated that a 10-micrometer receptive field looks like a quarter viewed from about 500 feet away. I don't think I can see a quarter at 500 feet; perhaps Nobel winners have sharper vision than the rest of us. In any case, we can think of these receptive fields as like the pixels on a monitor. The more densely packed the retinal ganglion cells, the sharper your vision.

SOME CONTEXT

In the early days of neuroscience—from, say, 1945 to 1980—the most exciting research centered on recording electrical signals. These included brain waves recorded from the scalp (an electroencephalogram, or EEG), distantly reflecting the electrical activity of the brain inside, and signals recorded by thin wires passed into the brain, reflecting the activity of single neurons. Recording the brain's electrical activity was the biggest game in town. (Molecular genetics—now the central engine of all biological science—was mostly biochemistry, and genetic engineering had not yet been invented.)

Needless to say, electrical signals picked up from single neurons are extremely small. That makes them susceptible to interference from all manner of other electromagnetic waves passing by, including police radios, television stations, and medical beepers. So we often recorded in "cages"—wire mesh boxes surrounding the test person or animal—to keep out the unwanted signals.

An even more basic way to keep out nuisance signals was to put something solid between the recording site and the source of the nuisance. How about several yards of dirt? Not a few laboratories were built in basements, or with copper screening set into their walls. (Apparatus has gotten better, and scientific interests have changed: we now record larger signals with better amplifiers, so it is no longer necessary to take such extreme measures.)

A typical research laboratory included three or four research groups, each headed by an independent leader. Each group would occupy a suite of three or four rooms, staffed by the leader (a professor) and three or four postdocs and technicians. Tiny offices for the professors would be tucked into the corners. A separate room would contain desks for the postdocs, or they would be squeezed in next to the apparatus. Animal quarters were usually a room located down the hall. On a visitor's first day, the pungent aroma of small mammals was overpowering. Fortunately, that smell faded and, after a few weeks, seemed to go away. The mice or rabbits, their bedding, and their feces were still there, but your olfactory system after a while tuned it out—a tribute to the blessed power of sensory habituation.

The gleaming flasks and bottles featured in movies were a minor presence in these laboratories. Instead, the dominant visual was of electronics—racks and racks of amplifiers, speakers, recording devices, and power supplies. If the lab was fortunate enough to have a computer, it would be the size of a refrigerator, have less power than my iPhone, and speak only a machine language. It was programmed by a specialist, using code not far removed from binary strings of 0's and 1's. Mixed with the odor of animals, alcohol, and ether was the smell of new wiring and warm metal as whirring fans cooled the circuitry.

Our equipment was precious. The big workhorse was the cathode ray oscilloscope, its dimly lit green screen the precursor to our present computer displays. We took pictures of the screen with a film camera. The oscilloscope had to be carefully calibrated, and it used vacuum tubes, like an ancient radio. My first task in the morning was to turn it on so that it would be warmed up by the time we wanted to start work. When I set up my own lab, its first scope cost $2,500 in 1970s dollars. Today you can buy a better one for $500.

STEPHEN KUFFLER

A pioneer of biological neuroscience, Stephen Kuffler helped create the discipline as we know it today. Kuffler did it not only by the example of his elegant publications, the lessons of quality he taught his students, and his personal charm, but by his skill in choosing students and colleagues, who today occupy a remarkably large fraction of the leadership in neuroscience nationwide. Those who knew him revere him. Machine shop technicians, secretaries, members of the scientific elite—everyone loved Stephen Kuffler.[5]

He was a slight, pixie-like man, and except for his enduring love of tennis, one would not have imagined that he was a champion in that sport during his youth. Born in 1913, he grew up in his family's mansion in Hungary. In his autobiography he calls it "a farm," but others have described it as a substantial estate, large enough to employ most of the people in the surrounding village. His early childhood seems to have been happy, although his family did need to flee Hungary for Austria during a short-lived Communist uprising in 1919. He was educated primarily at a Catholic boarding school, followed by medical school. Unhappily, his father suffered a catastrophic financial loss and died soon after, leaving young Stephen, then in his late teens, to fend largely for himself. Graduating from medical school in 1937, he once again had to flee, barely ahead of the German invasion of Austria. This time he fled in the reverse direction, from Austria to Hungary.

Via Trieste, he found his way to London, where he had friends. Not licensed to practice medicine in England, he moved again, this time to Australia, where he encountered John Eccles and Bernard Katz, both future titans of neurobiology, and began his life as a research scientist. During an intense period of work between 1939 and 1944, the three made fundamental discoveries about nerve conduction and the workings of the synapse.

Administrative bungles ended this charmed period, and the three left Australia, Kuffler taking his new Australian bride to Chicago. His growing reputation preceded him. After a couple of stops, he moved to Harvard, where he founded a department of neurobiology unlike any before it. If not the first academic department dedicated to neuroscience, it was surely one of the first few. The discipline at that time did not exist; the Society for Neuroscience had not yet been born. (My own membership number in the society is 000064, marking me as a real old-timer.) His department quickly became the leading one in North America. At first they accepted only a few students, which helped them create the family atmosphere for which the department became famous. A few years after its founding, I spent two years in Kuffler's neurobiology department as a visiting scientist.

The department's research standards were extremely high: it was elitist—not to say arrogant—and unashamed of it. In a way, it was the scientific equivalent of Ken Kesey's magic bus. If you were one of the family, you were on the bus. If you were not, they let you know that, too.

Inside the bus was a remarkable scientific environment. The growing group of scientists had fun. Kuffler himself was an inveterate jokester—one of the greatest un-stuffed shirts in all of Harvard. The people in his department were the Merry Pranksters of neuroscience, a strange and unusual mixture of rigidly disciplined, cautious, demanding, and playful. Seminars were supposed to be fun. Your talk had to be perfect yet effortless, and always leavened with humor.

The department ate lunch together, individuals or lab groups drifting in and out of the lunchroom as their experiments permitted. An important institution was the lunchtime seminars. As the department grew in fame, many visitors passed through Boston, and it was customary for them to be invited to give seminars. However, there were too many visitors to allow a seminar to be scheduled for each. The system of lunch talks was a solution. All that was required was for the inviter to write the speaker's name on the calendar posted on the door. There was no other announcement, and no formal welcome to the visitor. There was no official vetting of the speakers; they were solely the responsibility of the inviter. This was quality control enough—if you invited a bad speaker, you lost face. Worse, you risked getting your invitee flayed in public.

Over these lunches we heard an extraordinary amount of interesting science. Two or three such talks per week provided a truly comprehensive view of what was going on outside our lab. In practical terms, this gave us a competitive advantage: we learned very, very quickly—often before any publications appeared—of every new development. We were close to the leading edge, and proud to be there.

The department was not physically closed to outsiders—in a university at that time, that would have been gauche—but it might as well have been. No doors were locked, but scientists outside the department were not invited to attend any of the talks, or to eat in the lunchroom. Anyone who foolishly showed up was received with great coldness. Outsiders—that is, other Harvard scientists—naturally resented being frozen out, especially when they could look through the glass and see the neurobiology department laughing, playing, and doing wonderful science. But for Kuffler's gang, this was a magical era. It lasted until his death in 1980, after which the department began to fragment with remarkable speed. The Harvard Department of Neurobiology is still a marvelous place and a world leader in neuroscience, but those who participated during the glory days have never forgotten them.

(Honesty compels me to report that the experience was not entirely unmixed. The pressure to excel was great, sometimes overwhelming. One veteran of the early days told me that it was a great experience . . . and only took him two years of therapy to recover from. Also, the authoritarian scientific style sometimes led to errors.)

How did Steve Kuffler, a little guy overflowing with bad puns, have such an impact? His friends and students assembled a book of reminiscences after Steve's death. In it, Gunther Stent, a founding father of molecular biology, memorably called Steve "incorruptible." Many others spoke of Steve's scientific brilliance and integrity, but "incorruptible" says something more. It says that in his heart, there was something about Steve Kuffler that was pure.

He rejected pomposity and took any occasion to deflate it. Once, over a late beer in a nearby pub, accompanied by a couple of postdocs, a low-level professor, and Torsten Wiesel, a future Nobelist who had succeeded Kuffler as department chair, Wiesel was grumbling about his administrative chores. Steve remarked, with his usual slight smile but a very direct look, "If you want the glory, you have to do the work."

Steve gave me some direct advice at lunch one day, when he overheard me complaining to companions that the problem I was working on was not very general. (It seemed at the time to have implications only for the retina, not beyond.) Steve was sitting quietly a few seats away, eating his own lunch from a plastic carry-out box. He turned and, with that same direct look, said simply: "You learn something general by studying something specific."

CENTRAL AND PERIPHERAL VISION

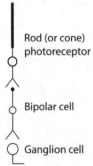

Rod (or cone) photoreceptor

Bipolar cell

Ganglion cell

To understand how the retina shapes the visual message, it's time to start thinking about how its neurons are arranged. The retina, far from being just a bank of photocells, contains five main classes of neurons, each of which does a different thing. The first of the retinal neurons are the rod and cone photoreceptors. (We say they come "early" in the visual process.)

These neurons detect light (rods for starlight and moonlight, cones for everything after dawn) and are the primary neurons in the retina to do so. The rods and cones make a synaptic output onto a type of interneuron called the bipolar cell, because it has—in contrast to some other retinal neurons—two clear poles, one for input and one for output. Bipolar cells get their inputs from the rods and cones and make their outputs onto the retinal ganglion cells, whose long axons bundle together to form the optic nerve. The retinal ganglion cells (which we say come "later" in the visual process) transmit to the brain all the information that the brain will ever have about the visual world.

We'll come in a bit to two other types of retinal neurons, which provide a lot of the fun of the retina. For now, however, photoreceptors, bipolar cells, and ganglion cells are the backbone of the retina, and their arrangement says everything about how sharply we can see.

Seeing a quarter 500 feet away is, of course, done with your sharpest vision. This is found at the center of your view, termed your fovea. Most people know that their peripheral vision is worse than their central vision, but they rarely recognize how profound the difference is. The sharp central vision of an ordinary person spans a circle of only about five degrees, equivalent to half the width of my hand at arm's length. Off to the side, visual acuity falls precipitously. In fact, a foot or two away from my hand (still at arm's length), I can barely count the fingers of an examiner. Eye doctors use a rough-and-ready phrase to characterize vision in this range: they say that the patient has "finger-counting acuity." In the pantheon of bad vision, the next step past finger-counting acuity is "hand movement only." Someone who has only finger-counting acuity is, in most US states, legally blind. In other words, we see sharply in our central field, but outside it we are pretty blind.

It is curious that we are so little aware of how bad our peripheral vision is. Somehow, as our eyes scan around the visual scene, we feel that we are seeing objects more sharply than we measurably are. This may be because we have a visual memory of the objects in the scene, having once fixed them with our central vision.

But our peripheral vision is far from useless. We use it in at least two different ways. First, it is very sensitive to *changes* in the peripheral visual scene. Something that suddenly appears, flashes, or moves calls our attention to that spot, and we are driven to direct our central vision to it.

A second thing we do with our peripheral vision is navigate. As we move through the world, coarse images of objects flow past us in our peripheral view. Even though we can't see fine detail, we are well able to make out large things: a doorway, a sofa, a refrigerator, the body of another person. This enables us to avoid obstacles and steer a straight path. This is dramatically demonstrated by unfortunate patients who suffer from a retinal disorder called macular degeneration. (Fifteen percent of white Americans will have macular degeneration if they reach the age of eighty.) For them, this means that some of the neurons in the fovea (where sharp central vision normally lives) degenerate. People with macular degeneration thus have pretty normal peripheral vision but poor or absent central vision. They are severely visually handicapped—unable to read, recognize faces, or watch television. But they can still walk around their living room; indeed, they can walk down a city sidewalk, albeit cautiously. Although they are legally blind because of their central vision handicap, an observer would not at first judge them to be as handicapped as they actually are.

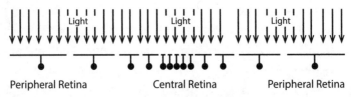

Peripheral Retina   Central Retina   Peripheral Retina

The cause of the difference between our central and peripheral vision is pretty straightforward. You can see from the figure above that your vision has a far higher pixel density in the central region than in the periphery. The pixels that matter in this case are the retinal ganglion cells, those last cells in the retina closest to the brain, whose axons form the optic nerve. In the figure, the black circles represent the individual ganglion cells, and the T-shaped structures above them show the area of the visual input sampled by each. (No other retinal cells are shown.) The size of these regions controls the size of the receptive fields of the cells. In the central retina, the ganglion cells are very numerous and densely packed. Here, each retinal ganglion cell's receptive field is very small. Out toward the periphery on either side of the center, the ganglion cells become scarcer, and the region of retina sampled by each (its receptive field) is larger. Larger sampling regions mean coarser pixels and lower visual acuity.

Why is so much of our eye's surface given over to such weak vision? Why not have those densely packed retinal ganglion cells all the way across the retina so that we could see sharply in both the center and at the edges of our vision? The virtue of the existing arrangement is once again efficiency. Retinal ganglion cells are expensive real estate. Not only do they take up space in the retina, but they take up space in the optic nerve: each retinal ganglion cell must send its individual axon into the nerve. The human optic nerve is normally about 4 millimeters in diameter. But if ganglion cells were packed everywhere at the same density as the central retina, the human optic nerve would be roughly the thickness of a garden hose. If nothing else, that would make it hard to swivel your eye in its socket.

And there would be no point in sending such an information-dense signal to the brain unless the brain could absorb the information. Try to imagine how the world would look if the whole visual scene were as sharp as the world around your fixation point. In principle it would be nice—the world would be as sharp as it is in a photograph (where you can see more of the image in your central vision). But what would you do with all that information? Could you think about all the information in the field at once?

A similar strategy is used by some smart bombs and other optically guided devices of war. Although the manufacturers are not eager to divulge the details, they use a coarse image to locate the region of interest, and then increase the pixel density over that region to resolve it clearly. The goal is the same as the brain's: navigate the world using the smallest possible amount of computational hardware.

WHY A HAWK CAN SEE LIKE A HAWK

A trip into another corner of the animal world may make this point clearer. Imagine a wheat field, not too long after harvest. The field is covered with wheat stubble and bits of cut stalks. At the end of the summer, the vegetation has been bleached to a sandy brown. Close to the ground, field mice forage for grains of wheat shaken off during the harvest. Overhead flies a hawk. Its wings barely move as it soars, gracefully maintaining an altitude of 20 feet. Suddenly it folds its wings and dives. It rises again, talons piercing a mouse's soft belly.

From that altitude, how could the hawk possibly have seen the mouse, an object only 2 inches long and hidden among the grasses? The mouse is small and dun-colored; the hawk is soaring along at speed. It's not for nothing that a person with very sharp vision is termed "hawkeyed." Many have looked for the basis of the hawk's performance, and some good observations have been made. One is that the cone photoreceptor cells—the very first neurons of the retina, the ones that are light-sensitive—are densely packed in a hawk, which is possible because its photoreceptors are thinner than those of most animals. Also, hawks have a large field of view—290 degrees in hawks versus 180 degrees for humans. Hawks have big eyes; the eye of a hawk takes up a surprisingly large proportion of its head, far more than in humans or other mammals. A large eye is a good thing:

remember that the larger the camera lens, the sharper the photograph. Professional photographers on the sidelines at a football game use lenses so large they can barely be held up by hand; the photographers prop them up with a support.

All this is to the good, but there is an element of bias in most of these discussions—the authors, mostly bird-lovers, *know* that hawks see well. They are seeking an excuse for hawks to see well, instead of objectively analyzing the hawk eye. Some of their explanations don't stand up to scrutiny. For example, cone photoreceptors in hawks are indeed small and densely packed, but they are only about 60 percent denser than in humans. Furthermore, the eye of a hawk is big (12 millimeters) for a small animal, but only half as big as that of a human (24 millimeters, about an inch). To be sure, humans have bigger heads, but that does not change the physics of light; when it comes to optics, humans are way ahead.

Finally, you can compare the hawk's visual resolution with that of humans.[6] With patience, you can train a hawk to choose between a target that leads it to a snack (say, a target with narrow stripes) and a target leading to a dead end (a target with wide stripes). You test how fine a stripe the bird can resolve. For the hawk species most often tested, the kestrel, visual resolution is in fact somewhat worse than in humans.

But wait! What happened to the hawk we started with, hunting a 2-inch, wheat-colored mouse above a stubble-filled wheat field? There is no doubt it sees better than we do—certainly better than me, in any case. How do we resolve this apparent contradiction?

I do not doubt the bird-watchers' observations. Here's what I think: the main reason for the hawk's superior performance is that it sees well almost everywhere, not just in central vision. The evidence for that is in the number and distribution of the retinal neurons. As noted above, a hawk's retinal cones cannot be much denser than mine—there is a limit to how tightly you can squeeze the cone photo-receptors together. But what is truly limiting for acuity is the packing of ganglion cells, not the packing of cones.

The key principle here is that *the resolution of any information-transmitting system is limited by the packing density of the least dense elements in the system.* In the retina—human or hawk—that would be the retinal ganglion cells, which represent only a few percent of the total neurons in any retina. As we have seen, in most animal species their density falls off dramatically at the retinal periphery. In hawks, the falloff is far less. In fact, hawks have many more total ganglion cells than people—about eight million per retina, compared to around one million for an average person. These are spread over the hawk's much smaller eye. And yes, the hawk's optic nerve is thick, but this is not troublesome because hawks do not move their eyes much, depending more on movements of the whole head.

What do hawks do with all those ganglion cells? For starters, they have two areas of central vision (two foveas, displaced laterally from each other), whereas we have only one. But the overall distribution is what's most important. The average human has only 1 percent of the density of ganglion cells in the far periphery as in central vision. But in hawks, the fall is slight—for the kestrel, the periphery has 75 percent the density of the central retina. These hawks are reported to have 15,000 ganglion cells per square millimeter in their peripheral retinas, while humans have only 500. Humans are virtually blind in the visual periphery, but hawks are far from it. The mouse has no place to hide, because the hawk can scan a swath of the field many meters wide with acute vision.

I asked a few paragraphs ago how a person could process the flood of information that would come if the sharpness of central human vision existed all across the visual field. It appears hawks are in al-most that situation, so how do they do it? We can only guess, but the answer seems to be that their brains contain a magnificent image-processing computer. A brain structure called the superior colliculus, which is also present in people, forms a strikingly large part of the hawk's brain. What its circuitry is doing can only be imagined—the subcortical visual circuitry of a human brain is lame by comparison. Someday, though, when we unravel the image processing that goes on in a bird's visual system, we will certainly learn some wonderful new ways to enhance images. Pay attention, Adobe—Photoshop may still have tricks to learn, from the birds.

3 | A Microprocessor in the Eye

A man bent over his guitar
A shearsman of sorts. The day was green.

They said, "You have a blue guitar,
You do not play things as they are."

The man replied, "Things as they are
Are changed upon the blue guitar."
—WALLACE STEVENS

So you now understand that you can see clearly where your retinal neurons are densely packed. But retinal ganglion cells are not all the same. They are not just photocells, like the "magic eye" that detects burglars in your house or prevents the elevator door from closing on you. They respond to different things, in precise analogy to the way different categories of touch neurons report on your skin. The visual world is fragmented, decomposed into a bunch of specialized signals. This initial step of image processing has consequences for seeing the sunrise, for dodging an onrushing automobile, for seeing your spouse, and for marveling at a Van Gogh.

IMAGE PROCESSING 1:
THE RETINA TAKES THE IMAGE APART

We'll start with the simplest kind of recoding, the difference between "sustained" and "transient" retinal ganglion cells. Some retinal ganglion cells generate a train of spikes mainly when the stimulus is first turned on; these are called transient cells. Others keep firing for as long as the stimulus is present; these are called sustained retinal ganglion cells. You may recall that this is exactly like the machinery that signals touches from your skin to the brain.

Superimposed on the sustained/transient distinction, there's another dimension that's important: among the sustained retinal ganglion cells, some fire a sustained train of action potentials for as long as the stimulus is on, while others are *inhibited* as long as the stimulus is on. The same is true with the transient cells. Thus we arrive at four different categories. Here are their names:

- Transient ON cells
- Transient OFF cells
- Sustained ON cells
- Sustained OFF cells

What does this mean for your vision? Imagine that you are the brain. Your task is to use the train of action potentials arriving via the optic nerve to deduce what event has happened in the world.

A transient cell responds mainly when a visual image first appears, then falls almost silent afterward. It is essentially a change detector. Clearly, you would not use the signals transmitted by a transient cell to recognize a face in a crowd. The face would disappear in an instant, after a few hundred milliseconds. You would not have time to register the configuration of its eyes, nose, mouth, et cetera. To see the face with a steady gaze, you, as the brain, would do better relying on the output of sustained cells. On the other hand, imagine that the shape of a pterodactyl suddenly swoops across your retina. That is something that the retina should report to you (the brain) as vividly and quickly as possible. This is what a transient cell is good for. Such a cell is silent most of the time. But its glory is to tell the brain about the sudden appearance of an object within its receptive field. Advertisers know that a flashing sign is more powerful than a steady one, and the transient cells explain why.

Some cells respond to brightness, and some cells to dimness. Brightness is easy to understand, but the concept of responding to dimness is a bit trickier. The two types of responses are termed ON responses and OFF responses.

Within transient cells, some respond when their receptive fields encounter an *increase* in brightness; these are transient ON cells. Some ganglion cells respond when a light goes off; these are transient OFF cells. Why are there ON cells and OFF cells? Remember that almost every visual object contains both a lighter edge and a darker edge. Consider a simple contour, dividing a light region from a dark region. Is this a light edge or a dark edge? It is both, and the retina signals both aspects to the brain. Imagine that you are intently reading this text until, cued perhaps by my narrative, your eye shifts so that it fixates exactly on the light/dark divider. What signals do your retinal ganglion cells send to the brain? A moment after your eyes alight, a subset of the cells covering the area to the left of the fixation point send a vigorous blast of action potentials to the brain; these are the transient ON retinal ganglion cells. They tell your brain that a lighter-than-average object has appeared within their receptive fields. At the same time, another set of retinal ganglion cells, covering the area to the right of the fixation point, are suddenly silenced; these are the transient OFF ganglion cells.

Yes, that's right: the brain gets two translations of the same message. The ON cells tell the brain something bright has appeared to the left of the divider. The OFF cells have given the brain the message in a different way. The OFF cells have also told the brain, "Things here have gotten brighter," but by firing *less*, not by firing faster.

A few tens of milliseconds later, the situation changes. The transient cells have done their job and fall pretty much silent. But how does the brain now know where the divider lies? That task is taken over by the sustained cells. The sustained ON cells set up a steady barrage of action potentials that continues for as long as you continue to fixate on the divider. And the sustained OFF cells are inhibited as long as your eyes continue to fixate on the divider. This contribution of the sustained cells is important: if your retina had only transient cells, the divider would become invisible a few tens of milliseconds after you first looked at it. You need the sustained cells for what we think of as acute vision—to resolve the fine details of the world, those that take a bit of time to inspect.

At the same time, the reverse signals are sent by the ganglion cells that survey the area to the *left* of the divider. Your transient OFF cells initially send a signal that there is an object darker than background somewhere within their receptive field, and transient ON cells signal the reverse. After a while, however, this signal too would fade, so the sustained cells take over, the sustained ON cells making clear to the brain "That dark thing's still there" in their own way, and the sustained OFF cells telling it "That dark thing's still there" by firing less and not more. Thus the retina has evolved to send a big signal when either a light object or a dark object first crosses the visual field—ON cells for a delicious fish gleaming in the dark water, OFF cells for the shadow of the owl, talons extended, silently gliding down upon you from above.

IMAGE PROCESSING 2:
IMPROVING ON THE REAL WORLD

Another important thing that the early retinal cells do is enhance the edges of your visual input. Note that ON cells and OFF cells do not transform the visual image; they simply pick a light or dark aspect to tell your brain about. Edge enhancement is different, because the original image is no longer faithfully transmitted to the brain. From the brain's point of view, it has been improved, in the sense that edges are where the action is, where the greatest amount of information lies.

That edges are important probably seems obvious. But edges embody a central principle that controls many, many aspects of vision. The pixels of the natural world are far from random. The natural visual world occurs in structures—lines, angles, curves, surfaces. What that means is that the occurrence of certain pixels is influenced by what's around them. A truly random visual world looks like television snow. Your visual system is arranged so that it emphasizes the structures where change is happening and downplays places in the image where not much is happening—the middle of the sky, the interior of a solid-colored surface.

Lateral inhibition is the mechanism by which the retina creates an enhanced response to edges.[1] This is a fundamental process for retinas and, as we shall see, for computer vision. Consider once again our simple edge. Remember that the middle regions of solid black or solid white do not carry much information. It is the point of transition—the edge—that carries the most information. Lateral inhibition increases the strength of responses of ganglion cells near an edge. Because of lateral inhibition, the difference between the signal that the brain receives from the black edge and from the white edge is greater than it would otherwise be. This is a prime example of the retina's selection of important features of the visual world for transmission to the brain.

It is just like the digital edge enhancement built into our phones and computers. As you are doubtless aware, a digital picture often comes with an associated command called "enhance contrast" or "sharpen edges." Turning it on makes the image seem crisper. Of course, there is no free lunch: the gray tones in the image may be lost. But sometimes it is worth it.

———

LATERAL INHIBITION IS omnipresent in sensory systems: it is found in vision, touch, and hearing and probably in smell and taste as well. It is seen in all mammalian and many invertebrate species, as might be expected for something useful enough that evolution invented it at the dawn of animal life: it was one of nature's first image-processing tricks. Why is lateral inhibition—edge enhancement—so useful to its owner?

For the answer, let's examine lateral inhibition as it is reflected in the signals sent by the retina's whole population of retinal ganglion cells. The sketch below shows the stages by which an image is transformed from a literally correct one as it falls upon the retinal surface (and is detected by the rod and cone photoreceptors) to the modified response sent to the brain by the retinal ganglion cells.

The top line shows a visual image. Half of the image is black and half white. The middle line represents the image being seen by a sheet of retinal ganglion cells. The line at the bottom shows the magnitude of the signal that ganglion cells send to the brain. Notice that right near the edge, the signal transmitted by each ganglion cell is boosted. The signal is increased on the light side. On the dark side, the response of the ganglion cell is decreased. The effect of this, from the point of view of the brain, is that the difference between light and dark—the signal that defines the presence of the edge—is increased.

For simplicity I have spoken here as though the retina contained only ON cells, when in fact around

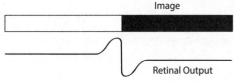

Image

Retinal Output

half of the ganglion cells are OFF cells. Their behavior is the reverse of the ON cells, but the net effect is the same—to increase the size of the differential signal in the neighborhood of an edge. I won't slog through the steps here. They are just the same as for the ON cells, in reverse.

Just for fun, here is something to reflect on: If the dark region of the stimulus is perfectly black, and the white region is perfectly white, does the black side of the edge look blacker than black, and the white side look whiter than white? If the dark region of the stimulus is perfectly black and the white region is perfectly white, the ON system and the OFF system are by definition at their limits—they can't go below zero or above 100 percent. But in the real world, usually all of the parts of an image have values somewhere in between—they're lighter or darker, not absolutely light or dark. When our visual system encounters a shift from lighter to darker, the lateral inhibition effect boosts the signal in the same way, heightening our perception of the contrast. This produces a well-known visual illusion called Mach bands: when two bands, one light and one dark, are right next to each other, we perceive a little extra-dark area at the edge of the dark band and a little extra-light area at the edge of the light band.

In summary, then, the retinal ganglion cells cover the retina with four basic types of detectors: transient ON, sustained ON, transient OFF, and sustained OFF. Each of the four is influenced by lateral inhibition so that its response near an edge is greater than its response in the middle of an unchanging field. But as we shall see in Chapter 4, the retina is even more complex—the retina is, to use the title of one research paper, "smarter than scientists believed."[2] But this took us a while to figure out. In the meantime, there were technical advances that let us look more closely at what the brain does with the information it gets from the retina.

DEL AMES: AN ISOLATED RETINA CAN STILL SEE

Many, many Nobel Prizes in biological science have been given, at least in part, for technical innovation. But most technical advances never make the front pages. One of these was by Del Ames, a brilliant scientist, a generous man, and my most important teacher.

Adelbert "Del" Ames III was descended from a New England family whose distinctions over the past few generations are too many to enumerate. The first Adelbert Ames, Del's grandfather, was a general in the Union Army. He is remembered today as the enlightened administrator of Mississippi after the Civil War, during Reconstruction. Del's father was a professor at Dartmouth and is famous for discovering the distorting effects of the surrounding environment on the perception of objects. We have "Ames rooms," in which perspective tricks make a person appear to grow or shrink as he or she moves around inside the room; you may have been in one at a carnival funhouse. (Del's father was also a skilled amateur sculptor. He sculpted the noble head of an Indian chief that in those days was found near the center of many New England towns as the logo of the Shawmut Bank.)

Del, a tall, lean Yankee of the robust Theodore Roosevelt stripe (he married Roosevelt's granddaughter), was an outdoorsman, a hunter and fisherman, physically bold. He lived to ninety-seven and was seen on cross-country skis at ninety-six. At Harvard College, he raced on the ski team, a sport he continued to age eighty-eight. He and some college friends built a glider, which in those days was usually launched by being towed by a car. After the initial boost, the pilot prayed for a rising air current. None of them knew how to fly, so Del made the first flight and taught the others. Years later Del helped his son David build a replica of an early stick-and-string glider. I helped them launch it, David racing down the slope of a hill near their Concord home before jumping aboard and taking the prone position of the pilot. It reached a height of perhaps a dozen feet, after which, alas, pilot error caused a stall and a wing-cracking crash.

World War II interrupted Ames's Harvard College career after three years, and he was rushed ahead to medical school. With some pride, he boasts that he never graduated from college. Harvard College refused to award him a diploma, giving him instead a certificate testifying he had "attended." Given his medical studies, talent in science, and demonstrated liking for cold weather, the army sent Del to Fairbanks, Alaska, to do research on winter fighting. There he experienced one of the coldest days on human record, $-82°F$. The army later put him to work studying the most effective way to rewarm airmen and sailors who had been exposed to extreme cold (as had also been infamously studied by the Nazi doctor Josef Mengele).

He and his colleagues learned some surprising things. The experiment began with a volunteer being immersed in a tub of ice water. Once his internal body temperature had dropped several degrees, the volunteer was fished out and warmed up. A 95-degree core temperature is very seriously cold—a further drop of a few degrees leads to violent shivering and painful constrictions of the veins in the hands and feet, and a few more degrees risks death. The point was to compare different methods of warming.

It turned out, Ames recounts, that some time-honored methods have their drawbacks. If you put a seriously chilled person in a warm room and give him a shot of brandy, his core temperature actually *falls.* The reason for this unhappy paradox is just a matter of physics. A very warm room may have an air temperature of 80°F. The core temperature of a severely chilled soldier may be in the neighborhood of 95°F. Alcohol causes the surface blood vessels of the body to dilate, allowing blood to flow easily to the surface, where it meets the 80-degree room. Although warm, that is still cooler than the person's 95-degree core temperature. Thus alcohol makes the chilled person's body a better heat exchanger, and this is a bad thing; the victim, who, though cold, is still warmer than the room, actually gives off heat to the environment, and his core temperature falls even more. (A better way is to put the person in a hot shower.)

After the war, Ames completed his medical training at Columbia. Then he returned to Harvard to have a try at research. He was not then a neurobiologist; he took a traineeship in the laboratory of Dr. Baird Hastings, chair of biological chemistry at Harvard Medical School. Ames, ever the independent thinker, started seeking a way to externalize nervous tissue for study. He wanted to get at the brain outside the head, so it would be easier to study. This was a radical idea, and Dr. Hastings told him it would never work, but he had been thinking about metabolism and could not think of anything prohibitively special about the metabolism of neurons.

At that time, clinical neurologists had a dominant voice in neuroscience, and they believed there was something ineffably fragile about brain neurons, such that they needed the support of their bony box to retain any function. These physicians had good reason for this belief. They knew that interference with the brain's nutrition for even a few minutes led to irreversible damage. If the heart stops, consciousness is lost within seconds, and there's a window of only a few minutes before the patient's brain (and thus the patient) dies or is reduced to a vegetative state.

In the frozen north, Ames had begun thinking about the metabolism of the brain. It turns out that neurons are indeed energetically demanding—more so than almost any other tissue of the body. The brain, which weighs only a couple of pounds, is estimated to use around 20 percent of the total energy supply of the body. Accordingly, it has a very rich circulation, the tissues nurtured by an exceedingly fine mesh of capillaries. From the capillaries to the neurons, passive diffusion carries nutrients and carries away cellular byproducts. But diffusion works well only over very short distances, so the brain has to be enmeshed in a vanishingly fine network of individual capillaries. We teach medical students that no neuron in the brain is more than two-tenths of a millimeter from a capillary. To put this in perspective, the density of the capillary meshwork is finer than the fibers in an ordinary bedsheet.

Ames wondered if there was a place in our central nervous system where neurons could be disentangled from their background of non-neural cells. He realized that the retina was such a place. Nonscientists rarely recognize that the central nervous system includes not only the brain itself but also the spinal cord *and the retina.* These three structures have the same embryological origin and the same kinds of neurons and supporting cells. Most important, all three are behind the blood-brain barrier, a system of coverings that creates a privileged chemical environment, separate from the rest of the body. Most of the neurons of the retina and spinal cord are bona fide brain cells. Taken one at a time, the neurons of the retina (other than the rods and cones) are indistinguishable, even to most neurobiologists, from neurons found elsewhere in the central nervous system.

But the retina has one problem not shared by other central nervous structures: it needs to detect light. If the retina were shot through with the usual mesh of arteries, veins, and capillaries, those structures (and the blood that they carry) would get in the way of the light. It would be as though we were looking at the world through a very dense window screen. The retina solves this in a clever way. The retina is a thin sheet of cells, rarely more than three-tenths of a millimeter thick. Because it is so thin, most of its circulation is within reach of diffusion from one side. The retina contains a handful of penetrating blood vessels that help nurture its farthest layers, but its major supply comes from a thick network of blood vessels located just outside the retina, at the back of the eye.

To Ames's further advantage, the retina of most mammals is not tightly attached to the cells on which it rests, so it's easy to detach. This is why retinal detachment is a risk. In people, a direct shot to the eye from a hockey puck or a hard-hit squash ball will do the job nicely. But the retina remains intact within itself—that's why, if the hockey player's retina is surgically reattached in time, his injured retina suffers only a little lasting damage.

Ames devised an artificial solution that copies in composition the fluid bathing the central nervous system (the cerebrospinal fluid). In the crucial experiments, he quickly removed the eye from a deeply anesthetized animal (later, without ever waking, the animal was euthanized), hemisected the eye, and gently teased the retina free, leaving it hanging from the optic nerve, which was then cut. Now the retina was floating free. It forms a delicately beautiful, almost transparent dome that is pale pink at rest, silver when bleached by light. Floating in a petri dish, it has the consistency of a piece of wet Kleenex, and its area is a little bigger than a teaspoon.

By almost any definition, isolated retinas are alive. They continue to consume oxygen and glucose. They synthesize new proteins. They give off metabolic waste products. Retinal neurons are electrically active. Over the next few years, Ames and his colleagues showed that isolated retinas behave as expected for brain tissue. Most important, they respond to light in very much the same way as retinas encountered in the eye of a living creature.

Over the next few years, Del's innovation spread throughout the discipline, and by 1980 almost no one studied retinas still contained in the animal. Indeed, it turned out that many other brain samples can survive for study outside the animal, if properly incubated. His special incubating solution, known universally as "Ames medium," is now sold by Sigma Aldrich Corporation, a leading supplier of laboratory chemicals.[3] By my very rough calculation, over the last forty years they have sold something like 300,000 liters of this solution, enough to float a US Navy frigate. (Ames never asked for any royalty or other payment for his contribution; when he later wanted to use Sigma's "Ames medium" in his own experiments, he had to pay for it.)

AN OPENING WEDGE

I was attracted by Ames's technology, and after graduate training I went to work with him at Harvard as a research associate. The experiments I did under his direction are as good an introduction as any to ground-level research on the biology of perception. Nothing fancy here, nothing prizewinning—just real science and a small but definitive advance that led on to other things.

Ames and I wanted to learn how the retina's circuits worked—to get inside the retina and find out how its internal neurons make retinal ganglion cells do what they do. We thought we could expose the many retinal synapses to drugs tailored to specific synaptic types. Basically, we wanted to give the system a precisely calibrated whack and see how it responded.

Our initial goal was to learn which of the different neurotransmitter types were used in the retina, but that was only a means to an end. Dozens of synapses link retinal neurons, and we wanted to take the system apart, tweak particular synapses, and see how the retina's signaling changed. For example, was there a particular neurotransmitter associated with ON responses and a different one with OFF responses? What were the neurotransmitters involved with the magical-seeming detection of the direction of moving stimuli? Would they lead us to the mechanism by which a handful of neurons in the retina figures out the direction in which something is moving? The basic experiment was pretty simple. Peering through the microscope, I gradually lowered the microelectrode we were using until its tip just touched the surface of the retina. Then, if I was lucky, the popping sound of a retinal ganglion cell would arise. (We detect the activity of neurons by amplifying the tiny signal picked up by a microelectrode.) If not, I would jiggle the electrode a little bit left or a little bit right, listening all the time for bigger and more stable trains of pops. Once a cell had been securely isolated, a crude air-cooled optical stimulator was swung into place. It generated little spots of light on the retina. I would flash light and listen for the characteristics of the response. After as good a characterization as possible, one or more test agents would be introduced via a side arm, and I would see if the response of the cell had changed. All this was done in near-darkness, using only dim red illumination—like the lights in a darkroom—to restrict accidental stimulation of the retina. The soundtrack was the hissing of the air supply and the steady crackle of the background neural noise.

And, oh yes, the whole room was heated to body temperature, 99°F. When Del designed the original experimental setup, he did not know what the important variables for keeping nervous tissue alive outside the body would be. It was quite plausible to him that temperature would be one of them. He therefore undertook to control it. He wanted the isolated retina to be at precisely its normal temperature. But the retina was being incubated in a flowing solution under a stream of oxygen, so how could he make sure the actual temperature in the incubation dish did not drift? Ever meticulous, Del decided simply to create an environment where *everything* was at 99°. Then everything—the retina, all the solutions, the flowing medium—would be at 99°, normal body temperature for a rabbit. He commissioned a little "warm room," heated by external controls to any temperature he chose. In winter, when the air is dry, being in a tiny 99° room for twelve hours at a stretch was not a hardship. It was less pleasant on a humid summer day. (One of my first actions after starting my own lab was to find another way to control the temperature.)

At the time, there were a few known neurotransmitter molecules. Intriguingly, however, all of them could be shown by crude chemical measurement to be present somewhere in the retina. We decided to use these transmitters as neuronal markers, to lead us toward the identities of the cells. Different functional types of neuron use different neurotransmitters, so we thought that identifying those would point out retinal cells that had specialized functions.

Acetylcholine was the longest-known and best-studied neurotransmitter. The retina was known to contain exceptionally high levels of acetylcholine: pound for pound, the retina contains more acetylcholine than almost any other structure of the nervous system. Preliminary experiments by Ames and Daniel Pollen had suggested that some retinal ganglion cells would be affected by acetylcholine. And because acetylcholine had been known for such a long time, there were lots of drugs that influence acetylcholine-mediated synapses.

I found out right away that lots of retinal ganglion cells were excited when acetylcholine or acetylcholine-like drugs were applied. The responses were consistent: cells excited by acetylcholine were also excited by agents that potentiate (that is, boost) acetylcholine's action. Some types of ganglion cells showed this coherent pattern of acetylcholine responsiveness and others did not, but it was not easy to see any coherent pattern among them. (I had thought that ON cells were more likely to be sensitive, but the division was far from perfect. I know now that my system for classifying the responses was too crude.)

I then proceeded to learn which cells contained the acetylcholine. Doing this was hard, and the project was made possible only by a long collaboration with my friend John Mills, artist of an arcane technique for freezing labeled acetylcholine in place. At the end of this road was a single finding: acetylcholine is contained in a single small group of amacrine cells. (Amacrine cells are retinal interneurons; they modify the firing of ganglion cells. I'll say more about them soon.) These later became famous as the "starburst" amacrine cells because their elegant, symmetrical shape reminded Ted Famiglietti, an imaginative neuroanatomist, of a firework of that name. As it happens, they turn out to be the principal driver of direction selectivity in retinal ganglion cells.

I completed a couple of other small projects during this era, but our main findings took seven years of almost uninterrupted work.

A PATH FORWARD

So we had shown that acetylcholine was a neurotransmitter in the retina and that it was contained in a small group of amacrine cells. But that was just one neurotransmitter, and we wanted to know the rest. We knew from biochemical experiments that the retina contained a lot of the other known neurotransmitter candidates—for example, lots of dopamine, famous in other parts of the brain for mediating reward, pleasure, and addiction. (No, I do not think that the retina is part of the pleasure system—dopamine has a different use in the retina.) Worldwide, a group of scientists, led notably by Berndt Ehinger in Sweden, took up the project of identifying which cells contained other neurotransmitters. As methodologies expanded, these studies got a whole lot easier, and in my own lab I joined in, with a particular twist.

My personal thought was that simply making a list of the retina's neurotransmitters was boring; what mattered was that they served as markers for specific cell types. What a handful of us did differently than most people was to insist on knowing both the complete shapes and the real numbers of the cells. We wanted to get away from the anecdotal style of classic anatomy, which some critics have termed "butterfly collecting." In this old-school style of research, you took a picture of a pretty example, added it to your collection, and that constituted your research.

I was interested in the numbers, connections, and whole arbors of neurons, and in particular the neurons that we would pick out from all the others because they contained specific neurotransmitters. (The "arbor" of a neuron is the tree of its axon and dendrites; since those are the elements that contact other neurons, the arbor of a neuron defines its possible connections.) What mattered to me were the *complete structures* and *numbers* of the cells—things you could use to build an irrefutable concept of how the retina's wiring was constructed, which in turn would tell us how it works.

The importance of this was driven home to me by a single stunning lecture given at a vision research conference by Heinz Wässle, a tall German about my age who was director of the Max Planck Institute for Brain Research in Frankfurt. The Max Planck Institutes are large laboratories, each headed by a single scientist. They are well funded by the German government, and there are not many of them. Max Planck directors are the cream of German science. At the time, Heinz was the youngest Max Planck director.

It was in this lecture, given at a conference hotel near the beaches of western Florida, that I first heard Heinz describe work he had recently done with Brian Boycott on ganglion cells.[4] They had figured out a way to stain two types of retinal ganglion cells—one a large ganglion cell present in relatively small numbers (which they called an alpha cell), and the other a smaller ganglion cell present in higher numbers (which they called a beta cell). Then, in collaboration with Australian researchers and his student Leo Peichl, Heinz showed that the anatomical shapes of the alpha and beta cells corresponded to their coding of the visual input. Alpha cells were the ON and OFF transient cells; beta cells were the ON and OFF sustained cells.

Why was this exciting news? First, it meant that the unique shape of a cell was telling us that it carried out its own private and unique role in the retina's functions. In fact, the more we learned, the surer we became that different shapes always meant different jobs in the retina's machinery—specific identifiers of the gears and wheels that make this machinery work. We could work backward from the shape to the neural microcircuitry, to the circuits that make that cell behave the way it does. So right before our eyes was a piece exposed of how of the retina worked, of the machinery that encodes the visual image.

The second reason it was exciting was the level of certainty that Heinz and Brian had achieved. Their anatomical studies gave us not just beautiful, butterfly-collector-type anecdotal pictures of "typical" alpha and beta cells but reproducible information about whole populations. Given that trustworthiness, their discovery of the stereotypy of shape was remarkable: just as maple trees have a typical branching pattern, different from that of oak trees, so do alpha cells have their own pattern, and beta cells theirs. These things are sometimes hard to see when you have only one instance of a cell type—a single maple tree—but given a whole batch of a single type, the consistent features jump out. With only a little practice, you can instantly recognize an alpha cell or a beta cell. And while Heinz had discovered two types, there must be others.

I had gone to the conference with a couple of good friends from Harvard, but they were from tangential scientific areas and had decided to skip Heinz's talk. I left them and their pitcher of beer at a rough-planked waterfront bar, the mild Gulf of Mexico breeze rustling the palm fronds overhead. When I returned, they were working on their second pitcher. I said to them, "I just heard something that will change the way we study neural circuits."

"What was that?" they asked eagerly.

I told them of Heinz's study and explained that we might soon be able to see whole cell populations, and could use their stereotypy of shape to identify them as cells with distinct functions, using real quantitative evidence rather than anecdotes. We could finally build something solid!

I could see their disappointment. They were thinking, "Anatomy? You must be kidding." But Heinz's talk had crystallized my thinking: I saw an algorithm, a bottom-up path that sooner or later had to tell us something important about how visual perception works.

As it turned out, our understanding of the organization of neurons in the retina—and in particular their great diversity—also presaged a change in understanding other structures of the central nervous system.

4 | Ghost Neurons

You Tiresias, if you know, know damn well,
or else you don't.*

—EZRA POUND

A QUIET REVOLUTION IN TWENTY-FIRST-CENTURY NEUROSCIENCE has involved the reinvention of the study of anatomy. Anatomy was thought by some to be dull—a pointless exercise, butterfly collecting. But the structure of the brain has always been important. The work of neuroscience's founder and patron saint, Santiago Ramón y Cajal, was based entirely on neuroanatomy. To be sure, memorizing the dozens of tracts and nuclei of the brain is among the least favorite tasks we set before medical students. Broadly, however, neuroanatomy—or, as it's now sometimes called, structural neurobiology—is where it's at: the brain is a connection machine,

---

* In a draft of *The Waste Land*, T. S. Eliot, speaking in the voice of the clairvoyant, Tiresias, had written, "Across her brain one half-formed thought may pass." His friend Pound did not like "may" and inserted this caustic marginal comment. It is good advice for scientists as well as poets—there's no point in almost knowing. (Pound's comment can be seen in *The Waste Land: A Facsimile and Transcript of the Original Drafts*, ed. V. Eliot [New York: Houghton Mifflin, 1971].)

and everything the brain does comes down ultimately to how its pieces connect with each other.

Around the year 2000, several advances added up to a leap forward in anatomical understanding of the brain. The first was a serious improvement in the resolution of microscopes, with the invention of something called a confocal microscope. (I'll show you one in action near the end of this book.) Another was an explosion of ways to visualize cellular components. The magical tools of molecular biology allow us now to create markers for even the smallest parts of the subcellular machinery, and confocal microscopes allow us to see it work. We can see things we never would have imagined we'd be able to see—cells in motion, swimming in their natural environments; clusters of cells that not only glow in the dark but glow different colors for different cell types. And these advances have allowed us to dream of the unimaginable: a census of *all* of the neurons in the brain—the brain's (or the retina's) complete "parts list," the first step toward unraveling the connections.

UNIDENTIFIED NEURONS

Heinz Wässle's work on ganglion cells, which I discovered at that Florida beachfront conference, suggested that we neurobiologists ought to take the retina apart from the bottom up. We could make a complete parts list, and only then try to figure out what the parts do. And as I've said, we had some great new tools to help us.

Immunocytochemistry, which came into its own around 1990, is a tool that lets us see the location of almost any protein molecule within a cell or tissue. Neurons that you might see in TV specials, glowing and spinning before your eyes, have most often been visualized using immunocytochemistry. The technique is basically easy, and the results are visually gorgeous.

It can also be frustrating. I could tell you a story about a commercial reagent that wasted a year of my laboratory's work. (In material terms, an unethical supply house cost American taxpayers about $300,000.) But we neurobiologists dived into it anyway: me and Julie Sandell, she first at Harvard and later at Boston University; Harvey Karten and Nick Brecha, pioneers in the technique itself; Berndt Ehinger in Sweden; Heinz Wässle and Leo Peichl in Germany; Dianna Redburn and Steve Massey in Texas. The New Zealander David Vaney took beautiful pictures through the microscope, revealing his first love: he retired young from science and started a new career as a photographer.

With the proper reagents, the method allowed anybody using a fluorescence microscope to light up all the retinal cells containing a specified target molecule. At low magnification you behold, against a dark background, a field of glowing stars. At higher magnifications, the detailed shape of a neuron becomes visible, its thin processes twisting across the retina or diving into it, the signature of that cell's pattern of connections with others. What reagent molecules would pick out which specific subtypes of retinal neuronal molecules? This was (and often still is) a matter of guesswork. The best probes so far have been synaptic neurotransmitters: dopamine, our old friend acetylcholine, serotonin, and the like, each of which was present in relatively small sets of retinal neurons. (Of course, neurons contain a lot more molecules than that, perhaps tens of thousands of them. But most of those are shared among many, many cell types, not only in the retina but also in the brain and the rest of the body; they do jobs like providing energy and maintaining cell structure. These were no use to us.)

Twenty or thirty scientific papers later, the group of us had accumulated a pretty good list of retinal cell types, roughly a dozen of them. Each of these cell types could be stained with great reliability, meaning that the whole population of that type of cell, all across the retina, could be seen separately from the retina's other neurons. We could measure their size, and we could count them—something that sounds trivial but is the root of real science, moving us away from butterfly collecting and the notion of a single, "typical" cell and toward a picture of what the cells do or do not contribute to vision. For example, some types of retinal neuron are present in very small absolute numbers but spread their dendrites huge distances across the retina. We could tell that such a population could never be involved in transmitting a high-resolution image. Too few cells gave too few pixels of vision; each cell gave input on too big a sample of the visual world to be useful for seeing sharply, because the image it provided to the brain would be dramatically undersampled and appear only as gigantic fuzzy blobs. Conversely, some tiny cells were present in large, densely packed numbers. We imagined them right away as parts of an important through-pathway, a high-resolution conduit from photoreceptors to brain, and that's exactly what they turned out to be.

So we and other labs were having fun making pretty pictures, and starting to guess how the retina's parts work. But after a while, we ran out of molecules to stain. Only a few molecules marked specific cell types, and nothing else we tried worked. Unfortunately, there was a much-ignored elephant in the room: most of the cells that we could identify were rare types. Now that we could see whole populations, we could tell that most of those cell types were widely spaced across the retina; there were wide-open regions where long rows of cells showed none of the marker molecules that we used for staining. If the retina were a kid's coloring book, we could color in about 20 percent of the parts, but the rest stayed black and white.

This was disconcerting. It seemed fatal to our attempts to understand how ganglion cell responses were created: If a large fraction of the potential actors are completely off our screens, how could we hope to understand the retina's computations—contrast enhancement, direction selectivity, and the like?

I'll admit that our greatest reason for trying to sort these cells out was simple curiosity. Suppose you were given an antique clock to repair, with no instruction manual. You would figure out before long that the pendulum was part of the timekeeping machinery. But what if there were a bunch of gleaming brass gears and wheels that just sat there, without any hint why they were necessary? Nature itself, the divine clockmaker, was teasing us.

The problem, for the retina and the rest of the central nervous system, is that without special stains, all neurons look alike. Our generic stains showed only the cell bodies, but it is the thin neural processes—the dendrites that reach out to receive inputs, the axons that send signals to other neurons—that make each type of neuron special. For that reason, the study of neuronal cell types had historically been anecdotal; we worked from examples we happened to stain, by methods largely dictated by chance and by hunches.

The retina, we thought, might be a place where we could do better. It has a geographic beginning and end; information flows through it in one direction. In contrast to many brain regions, we knew what the retina's task is. And it is spatially compact—the distance from photoreceptor cells to ganglion cells is only around a third of a millimeter. An achievable goal, we thought, was to create a map of *all* the cells in the retina, not just random examples. In modern terms, this listing of all of a structure's neurons would be called a neurome, by analogy with the genome, which is a listing of all of an animal's genes.

TRACKING DOWN THE GHOST NEURONS

But how to do it? We started at absolute ground level. Remarkably, there was only sketchy information on even the basic neuronal classes of the retina: photoreceptors, horizontal cells, bipolar cells, amacrine cells, and ganglion cells. To ordinary stains, the five cell types looked more or less alike, not much more distinctive than the little ovals in the sketches that illustrate this text. We knew that these big cell classes existed, and could guess roughly their proportions, but how to figure out what was actually there? The picture below shows roughly how the retina looked to us: we knew the identity of a few cells (the ones drawn darkly here, with their processes), but the rest—the open circles—were enigmas.

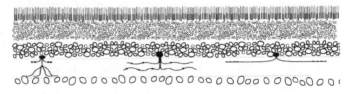

For advice, I went to Elio Raviola, a senior member of Harvard's Department of Neurobiology. Raviola is a scientist's scientist, master of all things neuroanatomical. I asked if electron microscopy, one of his many arts, could be used to see the differences between cell types. Of course, he said, but it would be incredibly labor-intensive: somebody would have to sit at a sectioning machine and cut tens of thousands of ultrathin sections through samples of the retina. Elio had better things to do, but he referred me to Enrica Strettoi, an Italian scientist who had recently finished a postdoc with him. Like Elio, Enrica is intelligent, friendly, outgoing, and absolutely uncompromising in her science. Together they had done a terrific project on neuronal connectivity in the retina, using serial-section electron microscopy. Enrica brought skill, discipline, and passion to the project, and she had the key insight that made our thing happen.

"We don't need to kill ourselves with high-resolution analysis," she said. "Instead of scrabbling among subtle differences, let's just identify the cells from their root definitions—the paths their processes take toward the retina's synaptic layers." On the scale of electron microscopy, these are big processes. In fact, Enrica pointed out, we could even see them using light microscopy, if we stretched it to its limits. That would take far fewer serial sections, because sections for light microscopy are ten times thicker than those required for electron microscopy, and they can cover much wider fields. So we approached test pieces of retina as three-dimensional solids (easy to imagine in this day of digital imagery, but hard to imagine before it) and cut a series of slices through each piece. Our goal was to definitively identify every last cell within our test samples.

We prepared the retinal tissues in Boston. Returning to Pisa, Enrica cut continuous series of sections through them and took a picture of every section. She then mailed the negatives to us, by regular international post, in Boston. (By the time this project was done, this step had blessedly been replaced by digital negatives and email.) The third member of this team was Rebecca Rockhill, my technician. Rebecca was a trouper: when I asked her to go into the darkroom and make a few thousand photographic prints, she did not protest. She emerged from the dark in only five weeks, rolling out stacks of 8.5-by-11-inch glossy prints still exuding the pungent fumes of photographic developer.

Sitting at a long table, we flipped through the stacks of images, identifying every cell. If you're sitting with us, you'd see, in picture number one, a bunch of neuronal cell bodies, irregular profiles cut through in various places. You pick a cell, any cell. You flip to picture number two and find the same cell, cut at a slightly different depth. You go to picture three, and so on, until you see a process exiting the cell body—an axon or a dendrite. You ask yourself, "Does this process go upward toward the photoreceptor cells or downward toward the ganglion cells?" In the next picture the direction of the process gets clearer. You keep on flipping, watching as the process extends itself into the inner synaptic layer or the outer. With a few more sections, the process shrinks and finally disappears. At that point you have traveled all the way through that neuron's cell body and its initial process. You can't follow the process much farther because after a while the dendrite or axon is too thin. But you can follow it far enough to know confidently if it is headed for the outer retina or the inner.

Definitively identifying the trajectories of a cell's processes means you can identify it as a bipolar, amacrine, or horizontal cell from the root definitions of the cell types (see the drawing on page 68): an amacrine cell sends processes only to the inner retina and a horizontal cell sends them only to the outer retina, but a bipolar cell sends processes to both.

So then we go back to picture one, and on the cell body we write, with a felt-tipped pen, "B" for bipolar, "A" for amacrine, or "H" for horizontal. Since this is the first cell of the series, we write "B1" or "A1" or "H1." And then we go on to the next cell.

I did some of them myself; others were done by summer students. (This may sound like a dull summer, but they seem not to have been permanently harmed. Two of them are now established neuroscientists.) Because every identified cell was named on the print, we had a bulletproof way to go back and check the decisions we had made. Thus, each step was rigorous: we accounted for every cell in our samples of the middle retina and learned the exact fraction of neurons that were horizontal cells, bipolar cells, and amacrine cells. The experiment left no room for error. That felt good.

Once we knew the large classes of neurons, we could ask a more focused question: How many amacrine cells are missing from our census? We started with amacrine cells because they were the largest class of inner retinal neurons and the least understood. Of all the amacrine cells that exist, how many are accounted for by the cell types that we know? Shockingly, the answer was that the known cells, added together, accounted for only 24 percent of the total amacrine cells.

Although it was great to have a clear answer, it was not the most encouraging one. Remember that our goal in all this—the reason we counted neurons late into the night—was to understand how the retina processes information: what messages it sends, via the retinal ganglion cells, to the brain. That is, how does the retina generate the first step in visual perception? If we want to know how a ganglion cell comes to send the signal that it does, we are out of luck when 76 percent of the potential inputs to that ganglion cell are invisible to us.

ENRICA STRETTOI

Enrica Strettoi, of the Istituto di Neuroscienze del CNR, is about five foot three, dresses with style, and laughs a lot. Though she does occasionally come to the lab in jeans, there is always something creative in her attire—no T-shirts for Enrica—and she wears heels most of the time. In my lasting image of Enrica, she is walking down a crowded street in Pisa at midsummer, perfectly composed, wearing an elegant white linen jacket and skirt plus pearls, her high heels easily navigating the uneven cobblestones.

She was born, raised, and still lives in Pisa, one of the first university towns in medieval Europe. Growing up, she lived above her mother's grocery store. Today, she and her family live in a gracious house in the suburbs surrounded by a walled garden, a farmhouse owned by the family of her husband, Luca, who was raised next door. Her mother and two daughters live nearby. One daughter is a physician; the other is a physician in training. In contrast to their ebullient mother, they are quiet and soft-spoken. They speak English well and carefully.

Enrica works hard. Like a good Italian, she loves cooking for her family on the weekend, but otherwise she hits the lab before anyone else and leaves late. She suffers fools politely, though not gladly. Her pattern is to work passionately for months and then take a substantial vacation with her family—in August at the shore or in the Italian Alps, and Christmas at home in Pisa. She is a devout Catholic. She and Luca are singers, performing in the local opera company. She signs emails to her friends, "Hugs from Enrica."

One of her projects aims to mitigate a common cause of blindness. A group of inherited diseases collectively called retinitis pigmentosa (RP) are caused by defective genes expressed in the photoreceptor cells of the retina. If you inherit one of these genes, your photoreceptor cells degenerate, and you go blind—sometimes within a few years of birth, sometimes over decades.

Enrica wondered if sensory input would affect the course of the degeneration. As a model system, she and her students studied mice that possess those same genes for photoreceptor degeneration. She raised one group of these mice in normal, boring cages and a matching group in cages that were, as she puts it, "full of toys": wooden blocks to climb, holes to hide in, and a running wheel for exercise. To her amazement, the retinas of the mice in this enriched environment degenerated far more slowly. Long story short, it turns out that the main beneficial effect came from the running wheel—from exercise—when combined with sensory stimulation.[1]

Exactly how this works is still not clear, and the field has not paid great attention to Enrica's finding. That exercise is good for you is hardly news—everyone knows exercise is a "wonder drug," slowing or preventing diseases from head to toe. But the slowing of retinal degeneration by exercise is a fact that any patient suffering from RP should know about. I don't like it that they don't. Since Enrica's science has always been impeccable, I trust that eventually these data will get more airtime. If I were starting to go blind from retinal degeneration, you had better believe I would take a nice run on my treadmill twice every day.

CORNERING THE AMACRINE CELLS

So how were we to get a grip on those 76 percent of amacrine cells we knew only as shadows? Trolling through the catalog of potential immunochemical markers was no longer showing us anything new. We needed some sort of unbiased method for identifying retinal neurons, one that could show us *all* the cell types.

The solution had three parts. First, we decided to use, instead of molecular markers, the shapes of the cells. The exquisite branching patterns of neuronal dendrites and axons have fascinated neuroscientists since the dawn of the discipline. Indeed, Cajal's elegant drawings of neuronal arbors have recently been the basis of an exhibit at MIT—an *art* exhibit. There have long been a few skeptics who maintain that the shapes of the cells are not particularly meaningful—a reflection of the cells' developmental history, perhaps, but not of their adult function. Yet the cell shapes are important for an unassailable reason: the shape of a neuron is a reflection of its synaptic connections.

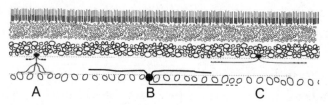

In the image above, three retinal neurons (A, B, and C) are viewed from the side, as though you were looking at a section through the retina. Cells A and C are amacrine cells (by definition: they send processes only into the inner retina, close to the ganglion cells). Notice that their dendrites spread out at different levels within the inner synaptic layer. This is crucial. The amacrine cell C cannot be in synaptic contact with ganglion cell B because they branch at different levels. In order to be in synaptic contact, cells have to touch each other, which these two cells do not.

The second crucial thing to notice is how widely the cells spread. Amacrine cell A and amacrine cell C must not carry out the same function. They must be different cell types because of their shape. A is a small cell, C is a large cell. Remember that the lateral spread of a retinal neuron determines how much of the visual field it looks at; again, some retinal neurons spread widely and others spread narrowly. Cells spreading widely are looking at a big piece of the world; cells spreading narrowly look at a small piece. Amacrine cells A and C are doing different things for vision, sending different messages to the retinal ganglion cell, and thus creating different messages for the ganglion cell to send to the brain.

But how were we to see the shapes of the cells in the first place? The images displayed above show the whole shape of the cell, something not visible under ordinary circumstances. It's easy to see the cell body—that's the home base of the cell, a storehouse of DNA and of the machinery that lets it make energy and build cell structure. But the processes—the axon and dendrites—are not easily visible. They are extremely thin and, more important, densely tangled with the processes of other cells. Even if you had, say, a stain that would stain all dendrites, you could never identify the dendrites of any individual neuron.

We needed a method that would make one neuron stand out from among all the others. Furthermore, it had to be a controllable method we could use to systematically sample the population of amacrine cells. The technique we settled on, photofilling, involved bathing the retina in a solution containing a photosensitive molecule that diffused throughout the retina into all of its neurons. We then focused a tiny spot of light, a spot smaller than an individual neuron, on a randomly chosen amacrine cell. Inside that cell, the spot of bright light set off a chain of reactions that caused fluorescent molecules to diffuse throughout the neuron that had been targeted, making it stand out against the background of all the millions of nonfluorescent cells.

It was a fussy method. For example, we could not take a picture of the fluorescent cell in the usual way, because the light we use to take the picture set off a fluorescence reaction in all the cells in the neighborhood. We solved that by buying an extraordinarily sensitive (and expensive) digital camera that let us take a picture in less than a tenth of a second, before the fluorescence had had a chance to spread throughout the neighborhood. The method worked better for small cells than big ones. But with practice, the operator—in this case Margaret MacNeil, a skilled postdoc—could very reliably work the reaction. In fact, when she aimed her spot at a randomly chosen cell, she succeeded in archiving an image of the dendritic arbor 94 percent of the time. This meant that a random sample of several hundred such images created a truly representative sample of the whole population of amacrine cells.

Remember the question with which we began this study: If the 24 percent of all amacrine cells that we can identify are the rare amacrine cells, what are the common amacrine cells? To our great surprise, the answer was that there *are* no common amacrine cells.

What does that mean? We had expected to find a few major players among the amacrine cells, surrounded by a supporting cast of specialized cells. Instead, the amacrine cells were distributed pretty evenly among a diverse set of cell types. Thus we had to assume that they were of equal importance in the processing of visual information. Our conclusion, which we published in a major journal (with only a little difficulty), was that there are twenty-nine different types of amacrine cell in the retina, each doing a separate job in the processing of the visual image.

Why is that little statement worthy of notice? It turns out that it was a pretty big clue to how the retina works. Why in the world does the retina need twenty-nine types of amacrine cells? The answer must be that there is much more processing in the retina than we had thought. Amacrine cells make a major output to the retinal ganglion cells, which are the last step before the retina sends its messages to the brain. If amacrine cells are so diverse, the messages must also be diverse. That was a step forward, a step toward understanding how visual perception works.

GHOST NEURONS CONTINUED: BIPOLAR CELLS

While we were working on amacrine cells, other folks were also
musing over the microcircuitry of
the retina. The most important gap
in the knowledge base involved bipolar cells. As you remember, bipolar cells get their synaptic inputs
from photoreceptor cells, send a
process through the retina, and syn-

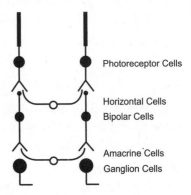

Photoreceptor Cells

Horizontal Cells
Bipolar Cells

Amacrine Cells
Ganglion Cells

apse upon amacrine and ganglion cells. They are an essential element
of the retina. If you took away all the amacrine cells in the retina,
it would still work, after a fashion; there would still be transient
ganglion cells and sustained cells, but the difference between them
would not be crisp, and there would be no direction-selective cells.
A person with no amacrine cells would still see, but her vision would
be fuzzy and slow. If you took away the bipolar cells, the retina could
mainly tell the brain that it was day versus night, a primitive function
carried out by a subset of intrinsically photosensitive ganglion cells.

Everybody has heard about scientific breakthroughs, where a new
observation or idea strikes like lightning from a clear blue sky and
transforms an area of science. It's far more common, though, that
science advances by accretion: as evidence accumulates, a possibility
turns into a probability, which finally turns into a fact. Such was the
case for our understanding of retinal bipolar cells.

The first systemic recordings from bipolar cells, by Akimichi
Kaneko, Frank Werblin, and John Dowling, seemed to show four
types: transient ON, transient OFF, sustained ON, and sustained
OFF. It would have been natural to think that four physical types
of bipolar cells would be the big drivers of those four types of ganglion cell response.

But for bipolar cells, as for amacrine cells, there was reason to suspect that not all of the types had yet been identified. By the mid-1990s, four or five labs were researching bipolar cells, and their estimates of the number of types ranged from four to nine. My own lab came late to this particular party. We figured that everyone's previous work gave us a good rough hint as to the organization of the bipolar cells. But this hint was to a large extent based on anecdotal evidence—"butterfly collecting," with a tiny sample of cells here and another tiny sample there. So we asked a different set of questions: First, are there bipolar cell types that previous techniques have failed to stain? Second, are any cell types predominant? Was there a boss cell with a few helpers? Or did all of the bipolar cells pitch in more or less equally?

To answer the question, we joined together with Elio Raviola. He had at one time done a series of gorgeous stains of the retina, but those slides were gathering dust in a bottom drawer in his lab.[2] A student had made a preliminary study of them, but the project had been dropped because Elio, ever the perfectionist, was acutely aware of the fickleness of his technique; he knew that there were probably bipolar cells that had not been stained.

My lab brought two things to the project. We contributed the photofilling technique, which gave us an unbiased—unfickle, all-inclusive—sample of bipolar cells. And we contributed Margaret MacNeil, who was by then a master at imaging neurons in three dimensions. Her images were our secret joy. Anatomists love beautiful pictures of neurons. We feel in the images of neurons something mystical, as though they show, right before our own eyes, a piece of the Truth.

We even had a third piece of information to guide a classification: images of bipolar cells that had been microinjected with a marker molecule after the electrical responses of the cells had been studied. This was the work of our friend Ray Dacheux at the University of Alabama.

Knowing how the cells actually responded to light was an invaluable complement because the responses of the cells turned out to be as distinctive as their shapes. The three methods—stains, photofilling, and microinjection—had different biases, making it hard to believe that any cell type would escape all three. Combining all this information gave us a lot of confidence in our result, which was that there were thirteen types of bipolar cells. Here they are, as drawn by Elio:

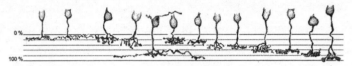

This picture emphasizes the defining feature of bipolar cell shapes—the depth of their axonal arbors. We found, as others had, that the biggest distinguishing feature of bipolar cells was the level in the retina's synaptic layer at which their axons branched. As you remember from the earlier amacrine cell drawing, the depth within the retina's inner synaptic layer determines which of the other players—amacrine and ganglion cells—that particular bipolar cell is in contact with.

We found that there was no predominant bipolar cell. Like amacrine cells, bipolar cells were distributed more or less equally among the individual types. Thus the retina has thirteen or so parallel pathways from photoreceptor cells to the inner retina, where they will meet the twenty-nine or so types of amacrine cell, and some large number of retinal ganglion cells, to create their final coding of the visual input for transmission to the brain.

As time passed, better stains were invented, and in a tour de force of immunostaining technique, Heinz Wässle and his students used them to make a new survey of bipolar cells. Their accounting was so precise and detailed that they could conclude that the numbers of the individual cell types added up exactly to the total number of bipolar cells in the retina (which Enrica and I had figured out). Wässle and his colleagues concluded that "the catalog of 11 cone bipolar cells and one rod bipolar cell is complete, and all major bipolar cell types of the mouse retina appear to have been discovered."[3] Even now, with fabulous new electron microscopic techniques and powerful molecular genetic markers, there have been only minor modifications to the catalogs of bipolar cells made by MacNeil and by the Wässle lab. The total number of types certainly lies between twelve and fifteen, depending on the criteria used.

These bipolar cells are the core of the retina. The dozen or so bipolar cell types represent the "primitives" of vision. Later evolutionary stages in the retina and in the brain can assemble messages sent by the bipolar cells in different combinations, modify them, emphasize some, and ignore others. But the brain cannot exceed the limits set by the bipolar cells. They are the elementary particles of visual perception.

BRIAN BOYCOTT

The most influential student of the retina in the late twentieth century was Brian Boycott, a Fellow of the Royal Society who never received an advanced degree.

I first met him in his lab at King's College in Drury Lane, London.[4] It was a dusty, seemingly empty place, wooden cabinets jammed with gear and ancient notebooks. Brian was by then a distinguished scientist. He was head of the Institute of Biophysics, where the famously unknown Rosalind Franklin had taken the X-ray diffraction images critical to Watson and Crick's discovery of the structure of DNA. He wore a plain shirt, no tie, and old slacks, and his belly drooped over his belt line. He smoked unfiltered cigarettes. I was a nervous, ambitious beginner, just beginning to be known outside my own university. We sat face-to-face on two lab stools. With very little hemming and hawing, Brian asked about my experiments, news of which had reached him through the grapevine. Our talk lasted until quitting time, the first of many long and smoke-filled conversations.

Brian Boycott was born in Croydon, England, in the winter of 1924. When Boycott was seven, his mother took him and fled his alcoholic father, leaving her and Brian with no financial support during the Great Depression. They stayed with friends for a few months, after which his mother obtained a low-paying job and was able to rent a single room.

By good fortune, Brian's absent father had once been a Freemason. This allowed the young boy to be taken in by one of the Masons' charity schools. It was a boarding school on the classic British model, providing him with food, shelter, and even clothing. He first enrolled at the age of eight; except for short vacations, he lived at the Masons' school for the rest of his youth.

Brian was beset with more than the usual complement of childhood diseases, but he does not seem to have been unhappy at school, though he excelled at nothing. He failed French and math, and barely squeaked through chemistry and physics. Rejected by the University of Cambridge, he enrolled at Birbeck College, a branch of the University of London dedicated to educating "the people of the artisan classes, in the evenings"—a night school. World War II had started, and this was the height of the Blitz; Birbeck College was firebombed, and so classes were held in the remains of a building whose surviving basement had been roofed over with corrugated iron. The University of London day students had been evacuated to comfortable safety in North Wales, but Boycott and his night-school classmates studied under the tin roofing, roasting in summer and deafened when it rained.

Scratching around for a day job, he landed one as an assistant in the zoology department's animal house, where among his other tasks was the job of cleaning cages. Perhaps his willingness to do this dirty job was a test, for he was soon moved to a more interesting position as a low-level technician in the physiology lab.

The physiology lab had, until the year before, been the province of Sir Henry Dale, a pioneer of synaptic biology. The lab retained Dale's structure and discipline. Brian spent his four college years working full-time in that lab, with classes at night. It was a fantastic experience for a budding biologist. In the animal house, Brian had rubbed shoulders with men of the British working class, whom he respected and liked. In Dale's lab, he rubbed shoulders with the scientific elite. In those days of small research groups, even the world-famous Dale probably employed not more than fifteen colleagues and technicians. I imagine that the brilliant young assistant must have been something of a pet, because he was soon given experiments to run on his own. One of them involved swinging dogs on a platform until they vomited. The research had been requested by the Royal Air Force, which for obvious reasons wanted to understand the basic biology of motion sickness. Brian did not tell me what they learned about nausea, but he did recount the story of a smart dog that learned to vomit at the sight of the apparatus, rendering himself ineligible for further study on that day. It is perhaps not coincidental that the neurobiology of Pavlovian conditioning was one of Brian's enduring interests.

Brian was later to write of his appreciation for that phase of his life, where he built and used his own apparatus, and found comradeship with a wide range of scientists. He wrote his first scientific paper during this time. It described a new method for measuring the amount of carbon dioxide accumulated in the rebreathing apparatus strapped on by underwater frogmen. Unfortunately, the military authorities stamped it as "secret"—even though the war by then was over—and Boycott's first paper was never published.

But he had evidently impressed his bosses. Following graduation from Birbeck, he gained a position as an assistant lecturer in zoology at University College. There his main task was teaching the laboratory section of the introductory course, a fundamentally menial job now given to graduate teaching assistants. He registered to do a PhD, but he put that on hold when he got a job as a research assistant to J. Z. Young, a world-famous zoologist, at a marine biology lab in Naples, Italy. Young's guidance was only that Boycott should see if he could learn something about learning by studying the brain of the octopus, the sea having been a fertile source of simple animals from which zoologists learned general principles.

Their research on the neural basis of learning was well received. Indeed, it was widely reported both in the popular press (learning in the octopus!) and across scientific disciplines. But a drawback was that Brian's famous boss cared more about pushing out the next paper than about Brian writing up a PhD thesis, which in the end never got done. In the meantime, however, young Boycott had achieved a substantial scientific reputation. In their inimitable fashion, the British authorities decided a PhD was not really necessary, and Brian was promoted to a faculty position. In later years, he was quick to correct anyone who addressed him as Dr. Boycott, proudly pointing out that his correct title was only Professor Boycott.

After returning from Naples to Britain, he got interested in other topics related to learning, among which was how the brain changes in squirrels when they hibernate. Partly because of the need for squirrels, he took a semester-long teaching position at Harvard, where ground squirrels were more readily available than in London.

At Harvard he met John Dowling, forging a partnership that was to result in a major advance. Dowling had begun a study of mammalian retinas using electron microscopy. But electron microscopy yields too high a magnification. Dowling had observed fascinating synaptic arrangements in the retina, but he could not know, from this high-magnification view, how to connect them to the cells they belonged to. Boycott was a master of whole-cell staining; from his studies of the octopus, he had learned a lot about unraveling neural circuits. Boycott and Dowling soon recognized their complementarity. Together they published a landmark study of the retina's connections, the fundamental basis of every understanding that has followed.

Brian had a penetrating view of the big picture, and a laser-like ability to detect the next set of crucial questions for understanding the retina. His own work, with a single technician, was on the fine structure of the retina, but perhaps his biggest contribution was to bring his breadth of view to those of us junior in the field. Brian spent several months each year at Wässle's lab in Frankfurt as advisor, critic, editor, and cheerful éminence grise. Wässle is a superb scientist and an efficient administrator, and he had the laboratory resources of the Max Planck Institutes at his disposal. So while Brian pondered, Heinz's lab sweated the details. Beneath their generational difference, Boycott and Wässle were similar men: large-spirited and moral, disciplined and tenacious.

Although Boycott lived simply, he was not straitlaced. He had a healthy interest in world events and the politics and sociology of science, as well as food and drink. He was the first in any group to suggest a trip to the pub. He had the highest possible scientific standards and was intensely critical, but he would talk to anybody, from janitor to postdoc to the most senior professor, about anything, whether casual topics, career issues, politics, or deep theory. If you had something to say, you got a respectful hearing. If he later decided that your standards were not up to his, he remained friendly but had little use for your views. He shunned large scientific meetings, deriding a popular one as a "gathering of sheep." His talks were little rehearsed, casual, and sometimes wandering—a far cry from the slick, TED-style presentations favored in the present era of easy graphics.

Brian reached out to any young scientist he found interesting. During his trips to the United States, he often came to stay with me in my little house near Boston. We would sit on my back porch late into the night, trying to sort out the retina's cell populations, gossiping about our friends, and drinking bourbon. After Brian died, the community of retina scientists voted to remember him with an award for scientific achievement, given once every two years when the clan gathers in the mountains of Vermont for its biennial meeting. The Boycott Prize is a cherished honor, as Brian was respected and loved; the awardee receives a certificate and a bottle of single-malt scotch.

5 | What the Eye Tells the Brain

> In one implementation, an encoder possessing
> a first set of learned visual knowledge primitives
> excludes visual information from an image prior
> to compression. . . . [Later a] decoder possessing
> an independently learned set of visual knowledge
> primitives synthesizes the excluded visual information
> into the image after decompression.
> —A US PATENT APPLICATION FOR
> A VISUAL RECOGNITION ALGORITHM

> It is impossible to *know* all of reality, but it is
> entirely possible to *understand* it.
> —BRIAN THOMPSON, *ZEN THINKING*

NOW THAT WE KNOW ABOUT THE RETINA'S MAIN PLAYERS, WE can start to understand what messages they forward to the rest of the nervous system.

So far I have told you a major principle: the resolution of all vision is controlled by the mosaic of retinal ganglion cells, in the same way that the density of pixels on a screen determines the screen's resolution. The more densely packed the retinal ganglion cells, the more sharply the person or animal can see.

And you also know some principles about the types of information that get transmitted to the retina: some ganglion cells respond mainly at the onset of light, some when the light disappears, some only transiently, others sustainedly.

There is much more known about retinal ganglion cells. Recent estimates suggest that in most mammals there are more than thirty different types of ganglion cells, each tuned to a different aspect of the visual stimulus. I'll now tell you about a few of the remaining types. The details are less important than the general concept. But the concept is important—remember, these signals are the final message that the retina sends to the brain. They contain all that the brain will ever know about the world of vision, because these are the only messages the brain gets from the eye.

"SMART" GANGLION CELLS

I've told you so far about garden-variety retinal ganglion cells: ON cells, OFF cells, sustained cells, transient cells. But there are others. The most famous of the "smart" ganglion cell types is directionally selective: it responds to movement of a stimulus in one direction, and not to movement of the same stimulus in the opposite direction. It responds, in other words, to the *direction* of movement itself—independent of the particular visual object. A direction-selective cell will respond to a light edge moving from left to right across its receptive field, but it also responds to a dark edge moving from left to right—in terms of physics, a very different stimulus. It does not care if the moving thing is a large object or a very tiny spot, as long as it moves from left to right. In the drawing, the dashed circle is the receptive field and the smaller circle is the visual stimulus. It responds to movements smaller than the size of the receptive field, no matter where the stimulus falls in space. The neural mechanism that can do this trick is something pretty special. I won't go into the details, but we all counted it a triumph when in 2015 my German friends solved the puzzle.

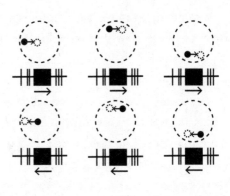

We know something of what this cell is good for: it helps control the position of our eyes when we move. Think about what happens when you look out the side window of a moving train or car. The scene is flowing past, and if it were possible for you to hold your eyes still, the image would be blurred by the movement. In fact, you cannot by force of will prevent your eyes from following the moving scene; they drift backward, then jump forward again to meet the moving image. If you doubt this, have someone watch your eyes while you look out the side window of a moving car or train.

Your retina's directionally selective neurons are largely responsible for this reflex. If you hold your eyes still, the image of the objects outside moves on your retina. The direction-selective neurons are activated: they tell the brain that the image is slipping, and which direction it is moving in. A nucleus in the brain receives that information and sends a precise message back out to the eye muscles, telling them how to contract to make the image stand still on the retina.

The importance of this is not mainly so we can ride in trains or cars. We have the same problem when we walk, and the motion is far more complicated. We walk—in effect, we bounce—from point to point. Our eyes must compensate for that exotic trajectory. Your directionally selective retinal ganglion cells help you compensate for these complexities by holding your gaze steady while you're walking.

Ask someone to move a page of large-font text from side to side in front of your eyes while you hold your gaze straight ahead, if you can. This is how the world would look without the mechanisms of image stabilization.

A second type of smart retinal ganglion cell is the local edge detector. It likes very slow movement of a tiny spot within its receptive field. Big stimuli do not excite this cell—indeed, nothing in the type of scene usually found at ground level excites it very well. William Levick, who discovered this type of cell in the rabbit retina, suggested that this is an evolutionary adaptation to the rabbit's life as a prey animal. For what it's worth, he pointed out that the tiny moving spot is just the stimulus that would be generated on the retina by a hawk circling slowly high in the sky. This type of ganglion cell is found in great numbers in ground-dwelling rodents, and the implication is that the retina of a rabbit or mouse needs to survey the sky for danger.

Earlier experimenters had encountered a neuron with the same selectivity in the retinas of frogs. They thought it was a bug detector, another reasonable supposition. But hawks eat frogs, too—so is it a hawk detector or bug detector? The truth is that we do not have a definitive use to propose for this cell. That will have to await an understanding of the whole process of perception—how the brain computes its final understanding from these inputs. In the meantime, though, these anthropocentric descriptions do convey something of these cells' behaviors and are easy to remember. And they remind us that evolution shaped these fancy neurons for a reason: to help the animal survive in that species' particular visual world.

A final example is a cell with a less-than-memorable name: the suppressed-by-contrast cell. Its response to light consists only of becoming silent when there is an edge in its receptive field. Mind, it has to be an *edge*—a big, diffuse light causes no response. And not only does the cell become silent, but it stays silent as long as there is an edge in the field. A distinguishing feature of the cell, also described by Levick, is that it has a high level of spontaneous firing in the absence of any stimulation. Because of that, the silencing is clear to any experimenter, and presumably would be clear to the brain.

I mention the suppressed-by-contrast cell because its usefulness to the animal is completely obscure. What this last type of cell does for vision is anybody's guess. And as I have said, there are many other cell types whose contribution to vision is not known. We know that these cell types exist, because we can see them: they have distinctive shapes, they cover the retina evenly, and they express distinctive sets of genes. Sometimes their functions seem obvious, especially when we give them names like "movement detector." What most of them do for vision, however, is far from clear.[1]

HOW THE FLEET OF GANGLION CELLS SURVEYS THE WORLD

Thus far, we have mainly been thinking about how individual ganglion cells signal to the brain one at a time. We have seen how the density of ganglion cells controls how sharply an animal can see—for example, in people and in hawks. Humans have about one million retinal ganglion cells. How do the different types work together to survey the visual scene?

Start by thinking about a single type of ganglion cell, reporting to the brain on whatever aspect of the scene it is tuned to. Obviously, you would like those cells to be present everywhere in the retina so that you did not have gaps in your vision for that aspect. Just as obviously, you don't want to have unnecessarily many ganglion cells. As we have seen, the retina tries hard to minimize the amount of neural hardware that's needed to do its job. Thus, the retina uses just enough ganglion cells of each type, and they "tile" the surface of the retina, as shown in the following drawing.

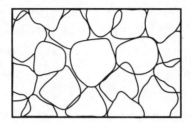

Now, what would happen when there is more than one type of retinal ganglion cell? Let's imagine a retina that contains three types, each indicated by different shading or outline in this drawing.

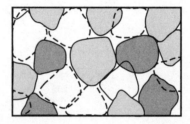

Of the three types, say one is a standard ON cell, one is direction selective, and the third is a suppressed-by-contrast cell. In the example as drawn, the retinal surface is covered by the three types of ganglion cells, but no one type completely covers the retinal surface. There are places where even though a ganglion cell is present, two of the three types are missing. If this were the way your retina really worked, the consequence would be gaps in your vision—more precisely, gaps in your ability to see the particular visual aspect reported by the absent type of cell. If, say, one of the three types was the retinal ganglion cell that detects motion, you wouldn't be able to see moving stimuli. (In fact, there are unfortunate people born with this particular lack. Their eyes are unsteady, oscillating rapidly from side to side.)

In actuality, each of the three types of retinal ganglion cells tiles the retina independently, as shown in the third diagram. That is, the

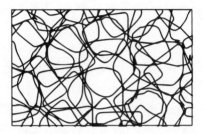

mosaic of each ganglion cell is superimposed on all of the others. If you stuck a pin through the retina at any point, you would hit the receptive fields of, in this example, all three kinds of retinal ganglion cells.

Since the visual world is projected as an image on your retina, this leads us to a powerful fact about vision: *Every point in the visual scene is reported to the brain by around thirty different analyzers (ganglion cells), each describing a different feature of the world at that point.*

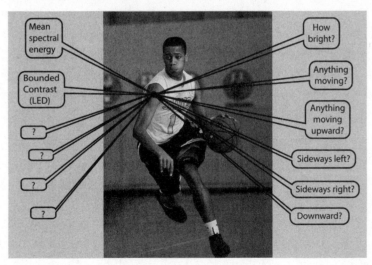

In this picture, I have captioned what the retina is signaling to the brain about a particular point in the image. The point in question is located at the edge of the basketball player's shoulder, indicated by the black dot. Different retinal neurons say different things about what they see at that point. For example, there is a set of neurons tuned to alert the brain if they see something moving to the right, to the left, up, or down. Another set of neurons are tuned to report where on the color spectrum each piece of the image falls. The local edge-detecting neurons, in contrast, send only a weak response, because in this location there are no edges (in this location they see a more or less uniform field, not a bug or a high-flying hawk). And finally, there are ganglion cell types that we know exist because we can see them by anatomy and by their differential gene expression, but we do not know what they tell the brain.

Another way to illustrate this point: What if you could see what a retina would see if some types of retinal ganglion cells were not present? In the image below, certain types are turned off (an approximation courtesy of Adobe Photoshop). As you can see, President Lincoln looks fuzzy if no edge-sensing ganglion cells are in play (left-hand photo), but he is harsh and unsubtle if only the edge-sensitive cells report on his face. When both are working, we see his familiar face in the usual detail.

Thirty different representations is a lot of information, which it is surely nice for the brain to have, but how can the brain absorb it all? How do these disparate signals, which have thoroughly deconstructed the original visual image, get put together into a single perception? What looks to us subjectively like a single image, a picture, has been split into a large series of different representations; you could think of them as parameters of the scene at that point. How these separate images get put back together again is one of the mysteries of perception, to which we shall return in the last chapters.

Before the 2000s, the retina was thought of as a simple neural system, with only a few major cell types. It was a shock to find that there were twenty-nine kinds of amacrine cell and thirteen types of bipolar cell. Indeed, there was pretty serious resistance to the idea. "You anatomists are just hopeless splitters," came a voice from the back row of the audience at one of my lectures. "You think every little twig makes a new cell type." But the evidence was incontrovertible: whenever there was a cell that we had a really good bead on—anatomists, biochemists, and physiologists all together—it turned out that cell carried out a distinct task in the retina's circuitry. A cell that has a different structure is always a cell that does a different job from the others.

There had been hints, mostly ignored, that the rest of the brain was just as complex. After publication of my lab's paper on the twenty-nine different amacrine cells, a respected neuroscientist calculated that there might be a thousand different cell types in the cortex—a number far above anyone's previous conception of the cortex, although the hints had always been there to see.[2] In the end, we realized that the retina was not simple . . . and that the rest of the nervous system was even more horrifyingly complex than we had imagined.

This picture shows many of the cell types present in a typical mammalian retina. It does not even show all of them, because cell types kept on being discovered after the picture was made. Even so, this image became the poster child for the complexity of the nervous system. This system introduced a thousand talks. The finding that there are so many cell types has changed our conception of how the brain works: instead of looking for a few players in different combinations, we now look for . . . what, a hundred different kinds of microcircuits?

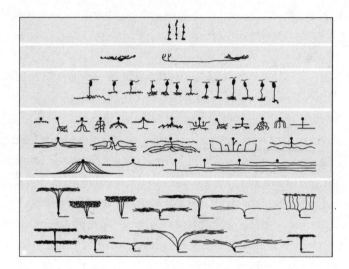

As it turned out, then, Steve Kuffler's pithy maxim was right: we learned something general by studying something specific. That is, we learned that the nervous system is far more diverse, and therefore computationally complex, than its canonical pictures had suggested—and we learned this by counting retinal neurons.

PART II
INTO THE WILD

LET US PAUSE FOR A MINUTE AND ASK WHERE ALL THIS PICK-AND-shovel science has gotten us. On the plus side, we have figured out a fundamental principle of visual processing: the visual image is decomposed by the retina into roughly thirty parallel streams, each reporting on its own specific feature of the visual world. In effect, our world is represented by thirty parameters, only a handful of which we currently understand. Different channels of information encode different features of the visual world—edges, lightness, movement, colors, and so on. Every point in the visual image is subject to the same array of thirty encodings. Furthermore, we'll soon see the remarkable transformation of these simple detectors as the information ascends to the cortex. For example, cells in the primary visual cortex respond to stimuli as precise as edges at particular orientations.

But how far does this take us toward our final goal of answering the question of how you recognize your child's face in a crowd, despite the hundreds of thousands of ways in which his image can arrive on your retinas?

From the primary visual cortex, located at the back of the brain, we peer like early seafaring explorers across a misty brain seascape, heartened by the few lands we know and have mapped, and driven to explore the many lands we do not know—the blank spaces in the map shown on page 106. How do we know the little we know about the vision areas on this map? Mostly from experiments in which we record from the neurons with microelectrodes, or scan their activity using imaging techniques. Doing these experiments we have found, as I will describe soon, that there are specific areas that seem to deal with pattern recognition, notably with the recognition of objects and especially faces. But our knowledge, like the map itself, comes in the form of islands—isolated facts with only a rudimentary story line connecting them.

In Part II, I'll describe this nascent story line, now being advanced by some leading scientists from neurobiology and computer science. This interpretation of the visual system is different from the textbook view, which postulated a hierarchy of evermore specific microcircuits, but without really saying what they are. It is one of our first serious attempts to connect more of the dots.

6 | Sensory Messages Enter the Brain

Ah, but to play man number one,
To drive the dagger in his heart,
To lay his brain upon the board
And pick the acrid colors out.
              —WALLACE STEVENS

S O FAR WE HAVE MAPPED THE OUTPUT OF THE RETINA TO THE BRAIN; but it is far from obvious what will happen next. Can we map the path of each of the thirty types of signal to their specific targets in the brain? It turns out that the answer is yes—for some but not all. We know where many go, and we have a clear idea what some do. In this chapter I'll describe a couple of the known targets, and end up at the visual cortex, the portal to object recognition.

WHAT HAPPENS FIRST

The output of the retina goes to two main places in the brain, carried by the axons of the retinal ganglion cells, which synapse upon neurons in these two different brain centers.[1] One of these is the lateral geniculate nucleus (LGN). The other is the superior colliculus.

Colliculus is Latin for "little hill." The old anatomists called it a little hill because it is a small bump on the back of the midbrain. It is the superior colliculus because it lies, logically enough, just above the inferior colliculus, which is concerned with hearing.

As best we can now see, the superior colliculus is concerned mainly with visual orienting. Signals from the retina go to the superior colliculus, and the superior colliculus tells us to pay attention to the particular place in the visual world where those signals came from. If you electrically stimulate a point on its colliculus, an animal shifts its eyes and the orientation of its head toward a particular point in the visual field. If an animal suffers damage to the superior colliculus, it appears to neglect part of the visual field: things in that part of the field never again attract the animal's attention.

Unfortunately, we cannot know the subjective experience of vision in the absence of a superior colliculus. This is something we would need to learn from experience reported by human patients. We can't know human experience because the superior colliculus lies only a centimeter or so above other brain centers that are critical for conscious life. It does not often happen that a human incurs damage that is restricted to the superior colliculus. There is almost invariably damage to the neighboring areas as well, in which case inattention to part of the visual field is the least of the person's troubles.

The superior colliculus contains tons of interesting-looking circuitry, with many interneurons, and lots of projections to and from other places in the brain. In fact, the colliculus has a layered structure, and some of the layers receive a map of auditory, not visual, space. It is still doing a kind of visual orienting, but in these layers, it is orienting the animal toward sounds, not sights. You could hear a sound with your eyes closed, and the superior colliculus would still direct your eyes to the location of the sound. It is quite common for visual and auditory cues to come from the same place in the visual world: perhaps the pterodactyl is cawing, or at least its wings make a flapping sound. In such a case, the visual and auditory inputs would synergize, giving you a strong and precise localization of the prehistoric bird.

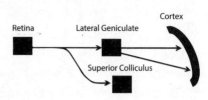

The lateral geniculate nucleus is the other major target of the optic nerve axons. (*Geniculate* is Latin for "knee-like." The structure of the lateral geniculate does have a sexy little bend.) The lateral geniculate is a brain "nucleus," composed of a cluster of neurons. Its neurons receive synaptic input from the axons of the optic nerve. Geniculate neurons send many of their own axons to the visual cortex. The lateral geniculate is the major way station en route to the visual cortex. Damage to the visual cortex, to the lateral geniculate, or to the pathway between them results in blindness in part of the visual field. The pathway from retina to lateral geniculate nucleus to visual cortex is the major pathway to conscious vision.

If the axons of the retinal ganglion cells synapse upon the neurons of the lateral geniculate nucleus, what are the visual responses of the geniculate neurons? In retrospect, the answer seems simple to predict—the responses should look like the responses of the retinal ganglion cells. And in fact that is what often happens. Recordings from the neurons of the lateral geniculate show that they also come in four main classes: transient ON, sustained ON, transient OFF, and sustained OFF, together with some of the "smart" visual analyzers. These signals are transmitted directly to the visual cortex.

However, as every graduate student learns, the lateral geniculate is not just a relay station. We make them learn this because we want them to understand how senseless it would be for nature to interpose a whole nucleus in between the retina and the cortex if nothing but dumb copying happened. We know, from anatomy, that the optic nerve is not the biggest input to the lateral geniculate nucleus. Amazingly, more axons (80 percent of the total input to the geniculate) run from the visual cortex to the lateral geniculate than run from the retina to the lateral geniculate. Although there have been a number of hypotheses, nobody knows for sure what this immense feedback circuit does. That's the way it is sometimes.

What does the lateral geniculate do for sure? A couple of talented labs have recorded simultaneously from ganglion cells in the retina *and* their individual target cells in the lateral geniculate—a good trick, I can assure you! In the cat and monkey, they found that the lateral geniculate neurons do copy, pretty much, the firing of the retinal neurons that drive them. (So too in the mouse, but with a subset of cells that receive very diverse inputs.)[2]

It's also clear that the lateral geniculate strengthens the process of edge enhancement: points of transition that are emphasized somewhat in the retina are emphasized even more in the lateral geniculate. It does this by connections between the through-pathway from the retina and local interneurons put there for just that sort of job. In fact, the edge enhancement is so strong that some neurons of the lateral geniculate are only active near an edge and hardly respond at all when stimulated by a large, smooth (i.e., not edge-containing) object.

Another thing that happens in the lateral geniculate is that the flow of information can be turned up or down by external events, specifically those relating to the overall level of excitation in the brain. When you sleep, the flow of information from retina to cortex is decreased. This is reasonable: it is a good deal like putting on the sleep mask that a flight attendant gives you during a night flight. A slightly more sophisticated function is that the throughput of the lateral geniculate is also turned up or down by selective attention. If you are concentrating on hearing, we think, you turn down your vision so that for the same stimulus fewer spikes are transmitted to the cortex by the LGN. The LGN has edited the message sent to the cortex.

AN ORDINARY DAY: RECORDING FROM THE BRAIN

We'll go on in the next section to talk about what happens next: where the output of the lateral geniculate goes and what happens to it. First, though, I would like to show you what it's like to dig for the kind of facts I've been talking about. Let me walk you through a day's work in the lab, to see what neuroscience looks and feels like when you do it.

I'll give you a postdoc's day of recording from single neurons. This is a composite; different labs have different schedules, layouts, and daily procedures. Some of the fabulous new techniques that we'll witness later dictate still another rhythm. But there are plenty of folks doing this type of recording just as I'll show here.

I do this since you would never otherwise read about it. The editors of scientific journals are not warm and fuzzy, as I imagine the editors of novels to be, nor do they care a whole lot about me personally. They are, at least in their day jobs, rigid sons of bitches whose idea of stylistic liberty is confined to how we write our names. Yet they are the ones who give most people their idea of how scientific discoveries are made. Scientific journals follow strict and formal stylistic rules, because their articles are designed to transmit precise information compactly, leaving no room for subjectivity and only the slightest sliver of (carefully labeled) opinion. Authors often get ten pages to describe a year's work. There is certainly no room for the scientists' personal experience of carrying out an experiment. So here's what a day was like.

We postdocs arrive in the lab at about 9:00 AM. The boss arrives a few minutes after nine, greets us, and goes to her desk. She reads and writes, mostly, keeping a distant eye on our activity. She is a full professor and a skilled experimenter—that's what got her her good job—but at this stage, we postdocs do the hands-on work. We call her only if we want to show off something new and cool . . . or if we have trouble with the apparatus.

Our lab consists of three rooms. The first is a general work area, a 20-by-20-foot square room. At the center is a surgical table, with a big surgical spotlight overhead. Along the back wall runs a black cut-stone lab bench. It holds a large sink, and glass-fronted cabinets run above it. In the cabinets are stored surgical instruments and the miscellaneous small impedimenta necessary for the work. On one sidewall is a bookcase shelving rows of scientific journals and a long row of laboratory notebooks, uniformly pale green with red sewn bindings. These document all of the previous work in the history of the lab; the books in current use are found at the right-hand end of this long row.

The actual recording takes place in a small dedicated room. It contains three tall vertical racks full of electronics. We turn them on at the beginning of the day, because they need time to warm up. After a few preparatory steps, our deeply anesthetized laboratory animal is placed in a holding apparatus, and the actual recording begins.

For this experiment, our goal is a simple one: we want to know how the lateral geniculate nucleus responds to inputs from the retina. Do the neurons of the LGN simply copy the neurons of the retina, or do they modify the signal as it passes through to the cortex? We have no preconceptions, no hypothesis. Each of us may privately have guesses, but they don't play a big part in our thinking. We are just taking a look around.

We monitor the LGN neurons by recording their electrical activity with microelectrodes. Even though the animal is deeply unconscious, its visual system still responds to stimuli—we can check that in modern times using techniques for painlessly recording from unanesthetized animals, or from people. The neurons of sensory systems fire spikes when their input exceeds a threshold.

A spike is an electrical event you can detect if you put a sensitive electrode next to the cell. The electrode must be very tiny, because you want to hear the signal from one cell and not its neighbors. The neuronal cell body, where it's best to record, has a diameter of around 5 to 30 micrometers (1 micrometer is 1/1,000 of a millimeter). In a nucleus like the lateral geniculate nucleus, other neurons are tightly packed against it on all sides. For that reason, your electrode must be pushed tightly against one particular neuron so that the signal from that neuron will be big and the signal from the more distant neurons will be small.

For this you need a microelectrode. (These days, your microelectrode usually comes from a commercial vendor. Until fairly recently, though, you created the tiny electrode by sharpening a long thin metal wire, like a hatpin, in an etching solution. Then you coated all of it except for the last one or two micrometers with an insulating material: plastic, varnish, or glass. You did this with great precision, monitoring the process through a microscope.) You attach the back end of the microelectrode to an amplifier and place the electrode in a fine micromanipulator.

Because it is so thin, you can push this long wire into the brain without causing much damage during the procedure. The neurons of the brain are not sensitive to pain. (When you get a headache, it comes from the surrounding tissues and the brain's blood vessels, not the neurons.) In human patients neurosurgeons are now using a therapeutic technique called deep brain stimulation, in which long, thin electrodes are inserted into the patient's brain. This is very often done while the patient is conscious so that she can report on her subjective experiences, and patients report no painful sensations as the electrode is pushed into the brain. It sounds gruesome, but it turns out to be a surprisingly innocuous procedure; thousands of these operations have been done, mostly to control the abnormal movements that come with Parkinson's disease.

Before we can search for a cell, however, we have to find the lateral geniculate nucleus. Remember that we can't see the lateral geniculate; it is deep within the brain, covered by the cerebral hemispheres. We use an apparatus called a stereotaxic frame to find it. This gives the position of the brain relative to landmarks on the skull. A chart specifies the three-dimensional coordinates of the structure we would like to reach, relative to those external landmarks. However, the charts are not perfect and there is some variation from individual to individual, both in animals and in human patients. For that reason, reaching your target structure is far from guaranteed.

A number of tries are usually necessary. We start with the microelectrode just above the brain, at the predicted X, Y location of the lateral geniculate nucleus. Slowly, using a micrometer screw, we drive it deeper until we reach the predicted location of the lateral geniculate nucleus. How do we know when we have succeeded? We know because our electrode starts to pick up signals that respond to light. We amplify the electrical potential generated around the electrode tip. We monitor those signals in two ways. First, we display them on an oscilloscope, which has a screen just like an old-fashioned television. This gives us a visual record of the activity. We see a horizontal line with the spikes sticking up out of the top and bottom of it. It is easier to see the screen if the screen is not brightly lit, so everything that follows is done in semidarkness.

We also amplify the signal and play it through an audio speaker, just like the speaker that you use to listen to music. Fortunately, the neural signals of the brain occupy the same band of frequencies as human hearing. In actual practice, the sounds of the cell are our main way to monitor its firing; the oscilloscope is a backup. Amplified, a single neuronal spike sounds like a short pop. A lot of spikes together—indicating we are not close enough to any particular cell— make a hissing noise. On an oscilloscope they look like grass, an unresolvedly dense thicket of tiny vertical blips. The formal name of this picture is "unresolved background activity." In practice we call it "grass" or "hash"—as in "Damn, we are losing this cell in the hash."

In the old days important events were recorded on magnetic tape or photographed directly from the oscilloscope screen; nowadays they are stored digitally.

At first we rarely hear an individual cell; instead, we see and hear a whole bunch of cells at once. That is because there is no one cell particularly close to the electrode; all of the cells are more or less equidistant, so their signals are more or less the same size. The first sign that the electrode has reached the LGN is that the activity of the cells starts to change. To test this, we use a low-tech instrument: a flashlight, the kind that takes two D batteries. We sweep it quickly past the animal's eye. When light is flashed in the eye, the unresolved grass thickens on the oscilloscope. The speaker emits a shushing noise. Flash, flash, flash becomes shush, shush, shush. At this point, we know we are near our target, and we slow the advance of our microelectrode down, until you'd barely notice we are moving it at all.

The operator gives the micrometer drive a tiny turn. The electrode advances. There are normally two of us postdocs working together. One of us monitors the screen for one blade growing up out of the grass; the other pushes the electrode and tries to stimulate the cell. Both of us listen for a faint popping to rise above the hissing sound of the hash. Every so often we just stop and wait for a minute or two; the brain sticks to the electrode, and a short wait allows the brain to gently slide back up the shaft. Another technique of microadvancing is to very lightly tap the table where the animal lies, the vibration transmitting itself ever so slightly to the electrode. Often the cell's signal grows out of the hash without any help: a faint popping is heard through the hash, and we advance the electrode the tiniest bit, seeking ever so delicately to close the distance between the tip of the electrode and the cell. Too fast an advance breaks the cell membrane, killing the cell. When that happens the cell fires an agonal scream of high-frequency spikes that quickly fade in size and frequency, like the scream of a victim falling off a building in a bad movie: *Aaaaiiiiiiieeeee!!!* But if we are skillful, we hear the song of the single cell: a steady drumbeat of spikes, with a burst when light is flashed in the eye.

By now it is usually early afternoon. Once we have isolated the signals of a single neuron, our task changes. Now we want to know: What does that neuron tell the brain about the visual scene? The experiment becomes a guessing game. A translucent plastic screen a yard square is positioned in front of the animal. On the screen is taped a thin piece of tracing paper. The animal's eye is focused on the screen. We monitor the activity of the cell mainly by ear. In the dark, it pops along at its own unstimulated rate. The task now is to see what kinds of lights, patterns, and motions cause the cell to increase its firing. We begin with a smaller flashlight—this time a penlight that generates a spot half an inch across. Rapidly we sweep the light around the paper (and thus across the retina), listening for increases in firing. Having roughly located the sensitive area—the receptive field—we switch to an even smaller spot, a couple of millimeters in diameter. Once again the spot is swept around the retina, generating a more precise localization of the receptive field. Carefully the edges of the receptive field are marked out in pencil on the paper. We will tape this paper in the lab notebook, as a record of the experiment.

Each cell identified in this way is named, using the date of the experiment and the order in which it was reached. So far we've only figured out what part of the cell's receptive field that particular cell is responsive to. We next test for direction selectivity: we sweep our spot back and forth across the responsive area, changing the direction of movement of the spot and testing various speeds and sizes of the spot. If the cell shows a preference for one direction of movement, that direction is carefully delineated and marked on the receptive field with an arrow. If the cell has no particular preference for movement, we conclude that it is likely to be a classic ganglion cell and proceed to see if it is ON or OFF, sustained or transient. Finally, we look for lateral inhibition. We do this with precisely timed, electronically driven spots, one placed in the center of the receptive field and one placed just outside it. We first record the response to the center spot alone, then to the neighboring spot alone, and finally to simultaneous stimulation of both. Almost always the paired stimulation gives a weaker response than stimulation of the center alone—lateral inhibition at work.

Beyond this, there are cells that do not show us right away what they care about. These are, therefore, none of the classic types. We know we have come to these when we sweep the spot around the screen and no particular area yields the high-frequency spikes that we see with the classic retinal ganglion cell. What do we do when we can't find out what the cell most likes? When we fail to find a strong driver of a cell, we are faced by a decision: Is the cell damaged—perhaps by our electrode—or have we just not found the specific stimulus this cell requires? If we never get a more specific response, we are frustratingly required to punt: the cell is recorded, but in the dismal category "unclassified."

Every cell is recorded by a brief note, handwritten in ink, in a sturdy lab book, whose sewn-in and numbered pages discourage anyone from taking a page out. If you make a mistake, the right thing is to lightly cross it out—no erasures or deletions—so that later readers can know there was an uncertainty.

Our notes are very simple. Take these from June 15, 1985, for example: "Cell 15/06/85-5, round receptive field, on center, surround inhibition. Cell 15/06/85-10, direction selective, preferred direction 7:00 to 1:00. Cell 15/06/85-14, responds weakly to diffuse light, cannot find anything better. Injured?" These notes, together with the receptive field drawings, recordings of the neuron's firing on magnetic tape, and oscilloscope photographs, form the main database of the study.

Lab notebooks occupy a sacred spot on their shelf, open to anyone. Faking data is hard—and in this case, at least, why would you do that? We had no expectation of our results, no theory to be tested—there would be no gain in making something up.

Lest I make this sound easy, note that there are many detours along the way. Sometimes there is equipment failure, or long periods of struggling to remove extraneous signals. (The amplification is very great and a microelectrode is a passable antenna; it is not uncommon to pick up the 60-cycle hum of the room's electric power, or even the soundtrack of a TV station. This must be defeated by rearranging the wiring or moving the shielding.) And then there are days when no cells are captured but the reason is not clear—probably a combination of various small mistakes by the experimenters. For all of these reasons, the average yield of successfully studied cells is around a half dozen per day. An experiment typically lasts from nine in the morning until mid-evening. If things are going especially well, we push late into the night, racking up as many cells as we can. Since we need a sample of several hundred cells to adequately characterize the pathway, the whole project takes many months. Many labs did this kind of work, and it forms the basis of much of what I'm telling you about the brain's coding of vision. The path to understanding perception is a slow one.

NEURONS THAT DETECT ORIENTED EDGES

The target of the LGN neurons—the primary visual cortex—does something extraordinary for a system that still includes only a few processing elements: it adds an additional wrinkle, one that caused huge excitement when it was first discovered in the 1960s. I can still remember the moment when I first heard about it, during the pre-meeting chitchat in a seminar room at Harvard. We were conditioned to think of neural responses as occurring in pretty simple chains. That a neuron so close to the periphery could detect the orientation of a line was a shock. It inspired a model of visual object recognition that held sway for several decades.

The core finding was that one type of cell in the cortex didn't respond very well to any kind of spot of light. What it needed was a long straight pattern of light—a line or edge. Even more amazing, it needed an edge aligned at a particular orientation. Its discoverers, David Hubel and Torsten Wiesel, termed it a "simple" cell (evidently because it was simpler than the one I'll tell you about next). Their diagram of the response of a simple cell to a line or edge of light is shown below.

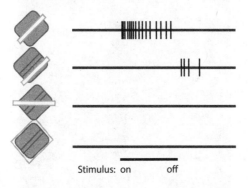

Stimulus: on off

In the picture, each vertical tick on the horizontal lines represents a single spike. This cell had a long narrow receptive field. I can tell this because if the stimulus was aligned with the receptive field, the cell fired a lot. If the stimulus lay a bit off to the side, the cell fired little or not at all. Importantly, if the stimulus was aligned at an angle away from that cell's optimum, there was no response at all. In other words, the cell was responsive to oriented lines or oriented edges that occur in the visual scene.

It also didn't respond much to uniform illumination, because the excitatory input and the inhibitory input would balance each other out. Same for an edge crossing the receptive field at other orientations: excitation and inhibition would cancel. Only when the scene contains a properly oriented edge did it fire. What they had found was an orientation-selective cell.

Why is orientation selectivity a good thing? The answer is that it once again reduces the amount of information getting transmitted to the next stages of visual processing (in this case, higher cortical areas). Importantly, it does this while retaining the key information important for identifying the object. Consider that the most important things in the visual stimulus are objects, and objects are defined by their edges. Tell the brain about the orientation of edges—and nothing else—and it can make a very good guess what the object is like.

Here the edges that would be reported by a cortical simple cell have been lifted from the dog's image and shown in isolation. You lose some of the richness of the original image, but it is still recognizably the same dog. These cortical neurons have taken the feature extraction process a big step past what happened in the retina. This particular feature, an oriented edge, gives the brain a sketch of the object—a simplified picture that costs less, in neuron currency, than transmitting the whole image. We can think of it as the difference between transmitting from your computer an image turned into vector graphics—most of the images that we ordinarily deal with—compared to transmitting a bitmap, in which every pixel is transmitted.

We know that bitmaps go very slowly, because they are an inefficient way to transmit information (even though the most complete way).

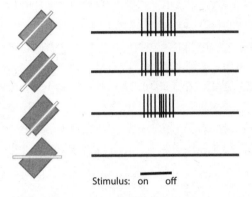

Stimulus: on off

The next type of cell, called a complex cell, also required the presence of an oriented edge, but the edge was not as tightly restricted to a single narrow spot on the retina. It responded as long as the edge had the correct orientation, but over a wider area. Again, vertical strokes indicate individual spikes of the cell, responding when the edge has the right slant and not responding to other angles.

To summarize, a simple cell is stimulated by an edge of light (or dark) located at a particular point in the visual field. A complex cell is sensitive to an oriented edge, just like a simple cell, but in this case the edge has a degree of freedom: as long as it has the correct orientation, it can appear anywhere within the receptive field, rather than being tied to a narrow region.

This was important because these cells can be said to represent an abstract concept, "lineness," somewhat unlocked from the exact visual stimulus. Even though the area is restricted to a certain extent, the cell surveys for lineness in a general region, not a particular spot. This takes us back to a problem mentioned at the start of the book: that we can recognize the letter A no matter where the A falls on our central retinas. In the 1960s the progression from non-oriented to simple to complex cells suggested a serial, hierarchical method of seeing more complex objects. As it turned out, that model did not really work. But the mechanism by which simple cells become complex cells did figure large in the invention of an important type of computer vision. More on all this later.

From here, neuroscience peers into a near-wilderness—the huge covering of the brain, its cortex. We really have only a kindergartner's understanding of the cortex. Fortunately, though, there are islands of knowledge—places where we have at least a rough idea of a cortical area's function. Even better, these areas are beginning to connect themselves into a landscape, a rough sketch of the brain's organization of perception.

7 | What Happens Next: Not One Cortex but Many

> There are known unknowns, that is to say we know
> there are some things we do not know. But there are
> also unknown unknowns—the ones we don't know we
> don't know. . . . [Those] tend to be the difficult ones.
> —DONALD RUMSFELD

NEUROLOGISTS AND NEUROSCIENTISTS BELIEVE (RIGHTLY OR wrongly) that the cortex is the seat of all those abilities that make us human—thinking, language, feeling. Although this may be an oversimplification, the visual cortex has been a major occupation of visual scientists. A huge breakthrough was the perfection in the late 1990s of painless ways to record from visual neurons while monkeys carry out learned behaviors so that we could correlate what we hear from sequences of spikes in the brain with the images we are showing to the monkey.

Up to now, when I have said "the cortex" I have meant the primary visual cortex—the main target of axons leaving the lateral geniculate nucleus. The primary visual cortex, V1 for short in the diagram below, is a patch of the brain's surface located all the way at the brain's back, taking up perhaps 15 percent of the brain's total surface area. As you would guess from the term "primary," there are more areas of visual cortex, named V2, V3, V4, and so on. There are also lots of areas that respond to visual stimuli but are not purely visual, or do entirely unrelated jobs and didn't get named as part of the visual system.

It turns out that the brain's visual areas are a patchwork. The various areas respond to objects in disparate ways and communicate with each other in ways we only poorly understand. Because the monkey cortex has been the focus of most work, from here on we will concentrate on monkeys, believing that, based on anatomy and lots of other evidence, human vision is not too different.

The brain shown here is labeled for specific subfunctions of vision. Each does a different part of the visual task, recognizing objects, detecting object motion, and so forth.

V1 is the primary visual cortex, the biggest target of visual information coming from the lateral geniculate nucleus. V2, V3, and V4 go progressively farther into the brain. You can start to think of them, more or less, as serial links in a chain of visual processing. But there

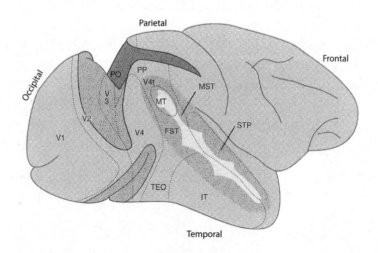

are a host of other areas, labeled in the diagram by initials that I'll translate where necessary.[1]

But first, what does all this brain real estate do for your vision? Using a very loose definition, some folks think that 30 percent of the brain is concerned with vision. There is way too much here to think about all at once (at least for me), so I'll begin by concentrating on two of the most studied cortical areas, one concerned with motion (area MT—middle temporal, near the center of the diagram) and a remarkable group of cortical patches within the inferior temporal lobe (IT), where cells respond to the presence of a face.

CORTICAL AREA MT: NATURAL MOTION SENSORS

Lots of hard work by smart people has detailed the response of cells within MT, shown on the diagram above. You can record from it, just like we postdocs did for the LGN, by putting an electrode near a neuron and then shining patterns on a screen and figuring out what you have to project on the screen to get the MT neuron to fire. It turns out that MT neurons have properties we have not encountered before.

First, MT neurons have receptive fields far larger than the receptive fields of neurons in the retina, or even the neurons of the primary visual cortex. They are four to ten times the size of visual receptive fields in V1, depending on how you make the measurement. This means MT cells can no longer be thought of as the pixels of vision. Their currency is something more abstract.

But MT neurons, most of them, do one very fancy trick: they are direction selective, somewhat like cells we encountered in the retina. Just like the retinal neurons, they tell the rest of the brain that something is moving in a certain direction. Unlike the retinal neurons, however, they don't tell the brain very much about where the moving thing lies, because of the size of their receptive fields. The moving thing could—judging only from the increased firing of the neuron—lie anywhere within a large region of the world. But they have other properties that make them useful.

Within their receptive field, MT neurons are pretty promiscuous. They respond to an ordinary big object if it is moving, but they also respond very well to patterns of small dots drifting across their receptive field. For reasons that I won't go into, the retinal neurons respond badly to this stimulus. You can tease an MT neuron by showing it a cloud of dots, some of which are moving in one direction and some in the opposite direction. (You write code to make your computer generate this pattern.) Under those circumstances, an MT cell will respond once it decides that a preponderance of dots are moving in its preferred direction.

So this cell tells its neighbors that there is something moving in its preferred direction, but it does not tell very much about what is doing the moving, nor very precisely where the moving thing is located in space. However, a truly remarkable thing that some MT neurons can do is to respond distinctively to the movements of objects that are partly occluded. An example is some MT neurons' response to an old-fashioned barber pole. A barber pole is a vertical cylinder that rotates on a horizontal axis. Importantly, all it does is rotate: its only true movement is a circular, horizontal one. On the cylinder is painted a pattern of slanting lines. When the barber pole spins, you see an upward (or downward, depending on the direction of the pole's spin) movement of lines on the pole's surface. This is of course not what is actually happening: any point on the surface of the barber pole is traversing a perfectly flat circle as the pole spins, and the upward movement is an optical illusion. Specialized lab tests show that many of the cells in MT report this same nonexistent direction of motion: they say that the stripes are moving up on the barber pole when in fact there is no objective upward movement.

Here is something even more remarkable. Many cells of MT are also sensitive to the distance of the visual object. The brain figures out distance by comparing where on the retinas of each of a pair of eyes an object falls. If an object is far away, the images on the two eyes fall very nearly in the same place on the retina. If the image is very close, the disparity is larger. Neurons in MT (and in some other areas) are sensitive to that difference in inputs from the two eyes. They are tuned to be sensitive only to objects lying at a specific distance from the animal. Remember here that neurons in MT have a preferred direction. Thus, a neuron in MT may respond only to objects located about 6 meters away and moving from left to right. MT neurons may be vague about stimulus size, but otherwise they are pretty damned specific.

And there is direct evidence that these cells participate in perception. William Newsome and colleagues at Stanford used their recording electrodes to apply tiny electrical stimuli to MT neurons that respond to a particular direction of movement, in awake, behaving monkeys. The monkeys were trained to report the direction of movement of objects. Stimulation of MT neurons was found to increase the perception of movement for real objects located in the cortical cell's receptive field.

Before going on, I would be remiss if I did not tell you that once information gets to the higher brain centers, almost everything seems to communicate with everything else. Here is a more detailed map of the connections in just the motion pathway—that is, the pathway centered around MT. You can see that this is, in brief, a mess: it looks like everything is connected to everything. And most of those connections have unknown functions—they are far less understood than, say, the neural pathways through the retina and to the lateral geniculate nucleus and primary visual cortex (V1).

The neurons of MT do very sophisticated analysis of the movements occurring in the natural visual scene, but when taken by themselves they contain a fair degree of indeterminacy. In short, then, we know some of the things that these cells do, but we do not know why they do those things—how their analysis of the visual scene contributes to our finally unified perception. We'll leave area MT for now, as it is part of a different pathway from the one that does object recognition.

PATCHES OF CORTEX THAT DETECT FACES

As we have seen, the first stop for visual information in the cortical brain centers is the primary visual cortex, V1. Then come V2, V3, and V4. Roughly speaking, information does flow from V1 to the temporal lobe through these areas, but they are poorly understood; despite lots of work, the sages have not come up with a pithy description of their function. (To jump way ahead, this may be because they are hidden layers in a nerve net, whose function depends on what the nerve net has learned. We'll come back to that in Chapters 10 and 11.) I'll only sketch them here to complete your picture of the brain's visual processing.

The clearest distinction between V1 and V2 is that V2 neurons have receptive fields that are larger than those of V1 neurons, and more of them have complex receptive fields. As we have seen, V1 contains a greater number of simple neurons, which are tightly tuned to a fairly narrow set of stimulus features (i.e., oriented lines in a particular place), and V2 has a greater number of complex cells, whose responses are less restricted as to place (i.e., oriented lines less tightly tied to retinal location). Even that, however, is relative: many neurons in V1 also have complex receptive fields.

Neurons in V3 have a mixture of properties. Almost all cells show orientation selectivity, but directional selectivity and color selectivity are quite common as well. V4 was once considered to be a "color center," but then we learned that neurons in V4 can have a variety of selectivities; V4 also contains cells selective for orientation, motion, and depth. So we can't give a single description of what these areas do. Although cortical areas V1, V2, V3, and V4 are sometimes treated as a hierarchy, the only thing that's really certain is that we progress from simpler properties to more complex ones.

But further forward along the temporal lobe lie systems of neurons with remarkable selectivity for objects. These include a series of areas (termed "patches") containing neurons selective for faces. Regions between the face patches are sensitive to other features of the visual scene, so the temporal lobe appears to be a checkerboard of regions taking care of different things. The existence of face-selective neurons was first reported in the late 1970s by Charles Gross and his colleagues at Princeton. They encountered neurons in the inferior temporal lobe that responded with great selectivity to particular objects, hands, and faces.

Face-selective cells seemed to most of us improbably specific. And in the space of the whole temporal lobe these neurons were not terribly numerous, so Gross's reports were greeted with some skepticism. If you study the temporal lobe with a microelectrode, as Gross did, you are at the mercy of chance: you have to study the kind of cells that happen to lie underneath your electrode and you don't get to study very many of them, because recording is a slow business and because the patches are patches—they cover only a small fraction of the temporal lobe's surface. It was only with the advent of brain scans that the face patches could clearly be seen and their existence verified.

Cellular neuroscientists—like Charles Gross and me—were initially scornful of MRI scanning as a neurobiological method. In contrast with the precision of microelectrodes, MRI scans only show gross areas of the brain, and in their early days, they gave pretty low resolution. To do imaging reliably requires skill and caution. Signals received by an MRI are small and subject to many kinds of interference. For that reason, the images are highly processed. Small deviations in processing can easily yield false results (not a few of which appear in pop neuroscience articles). But the machines have gotten better, and they have two advantages: first, they are completely noninvasive; second, they show activity in much of the brain at once, albeit at a resolution much, much lower than that of microelectrodes.

MRI images can be collected from living, conscious people or animals without any harm. This is because when an area of the brain is working, it needs more energy and thus more blood flow, and this is what fMRI imaging (functional MRI) detects. fMRI reveals to the experimenter which part of the brain is active at any given time. Subjects can be instructed to carry out various kinds of mental work, can be shown pictures, or can be played sounds; the resulting fMRI images will show which brain areas are involved in that particular task.

A turning point occurred when several laboratories, notably Nancy Kanwisher's at MIT, imaged the brains of human subjects while the person looked at pictures of various things. Lo and behold, specific small patches of the temporal lobe lit up when the subject saw a face. And the regions of the lobe were reliably similar from person to person—this was a reproducible biological event, not an artifact of the technique. The patchiness explained why Gross and earlier experimenters had trouble reliably demonstrating face cells: the microelectrode needed to be in just the right place.

In people and monkeys there are six such patches, arranged along the surface of the temporal lobe from posterior (left in this picture), nearest to the primary visual cortex, in a crooked line toward the front end of the lobe.

The exact locations of the patches are somewhat variable, and some of them are underneath the brain, so most images like the one above, which is a side view of a scan from the brain of a small monkey, can't show all six at once. In each patch, however, a high  fraction of all the cells are selectively sensitive to faces—human faces, monkey faces, cartoon faces, doll faces.

Then Doris Tsao, Margaret Livingstone, and their students learned how to direct microelectrodes to each face patch. Remember, the face patches are defined by neural activity, not neural structure: there are no completely reliable surface landmarks to show where a face patch might lie. These groups figured out a way to image the face patches and then reproducibly record from the identified region of the temporal lobe with microelectrodes. This was one hell of a brave experiment to undertake—anyone who has ever placed their head inside an fMRI scanner can imagine the task of getting a monkey to lie still in one.

It turned out that all face patches contain neurons with very large receptive fields. They survey regions of visual space far larger than the neurons of the retina, the lateral geniculate nucleus, or V1. Within that large area, the cell reports when it sees a face. But there are differences between the various face patches. In the most posterior patches—the closest to V1—the responses of the neurons are view-specific. This means that the face to which they respond must always point in the same direction to be recognized. If the face is your grandmother's, she must be looking toward your left shoulder.

Another area appears to respond to a given face or to its mirror image. In other words, this cell has made a step toward a key element of object recognition, freeing itself from the requirement for a face in a place; the response of the cell is partially view-invariant. And the last patch, closest to the frontal lobe, contains neurons that are truly view-invariant: they recognize a face no matter what angle it is seen from. Thus, a leading hypothesis is that the six patches represent a hierarchy in which the earliest patches are more strictly tied to the retinal image and the most anterior patches are less so.

The patches work together as a system. This evidence comes from experiments in which the patches that recognize faces were electrically stimulated. Monkeys were trained to recognize faces, and then the experimenters stimulated one patch or the other with very fine microelectrodes. First, the patches turn out to be linked: stimulation of one face patch causes activity in the others. Second, electrical stimulation that disrupts the normal neuronal activity makes the monkey less able to discriminate faces. This confirms that the face neurons are actually used to recognize faces.

We all know how a face looks, but what, precisely, does it mean when we say that a cell "recognizes" faces? The quality of "faceness" can be broken down, as you probably already suspect, into different elements. One is the presence of two eyes. A second is the presence below them of a more or less vertically oriented line—a nose. Better still is two eyes, a nose, an oval below the nose, and so on. Experimenters have added and subtracted such features from real or artificial faces and found that the cells' responses grow progressively weaker if some features are missing. Thus a face cell responds weakly to the face on the upper left of this image and strongly to the face

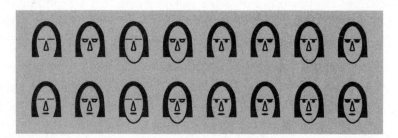

at the lower right—but it does respond somewhat to all the images in the panel, including the image that has only a few facial features.

It has been proposed that what these cells are doing is measuring a selection of parameters of the face and using joint analysis of the parameters to decide whether something is a face. For example, Tsao and her colleagues have studied a face-selective cell that responded particularly well to four things in a face: the ratio of height to width, the inter-eye distance, the location of the eyes, and the size of the pupils. No one of these defined, for that cell, the presence of a face, but taken together they were able to convince the cell it was seeing a face.

Each cell of a given face patch responds best to specific parts of the face. By making cartoon faces and then fragmenting them into different aspects, experimenters could learn that some cells are most sensitive to the aspect ratio of the face (long, thin faces versus round faces), others to the distance between the eyes, and so on. The cells don't so much "recognize" faces as measure parameters of the stimulus that is potentially a face, and then somehow sum them up to decide whether the stimulus is a face.[2]

How do the cells do this task? How do they become selectively sensitive to faces? My belief, and that of some others, is that they *learn*. The theory depends on a single critical fact: there is great plasticity in the neural wiring of sensory systems. This is worthy of a chapter unto itself.

8 | The Malleable Senses

Subtle, clever brain, wiser than I am,
by what devious means do you contrive
to remain idle? Teach me, O master.
— William Carlos Williams

THE EARLIEST THINKERS KNEW THAT THE SENSES ARE "MALLEABLE." A commonplace observation is that people deprived of one sense may have a compensatory increase in the others—blind people have heightened senses of hearing and touch. Furthermore, experiments on perceptual learning, like the ones we will discuss in Chapter 9, show that a person's sensory acuity can achieve major improvement with practice. But a skeptic could say that this was just a matter of attention, concentration, and practice at the task, rather than a true sensory improvement. We had to use modern methodologies to conclusively prove that the circuits of the brain neurons were physically changed.

DAMAGED SENSES REWIRE THEMSELVES

The term "brain plasticity" means the ability of the brain to reorganize its circuits. The paradigmatic early experiment went like this: Take a rat that has been deprived of vision since birth—let's say because of damage to both retinas. When the rat grows up, you train that rat to run a maze. Then you damage the visual cortex slightly. You ask the rat to run the maze again. (The rat is happy to do so, because he gets food at the end of the maze.) You compare how fast the rat runs the maze before the operation and after. In principle, damaging the visual cortex should not do anything to the maze-running ability of that blind rat. But the finding is that the rat's performance does get worse, suggesting that the visual cortex in the blind rat was contributing something to maze-running, although we do not know what it was. This experiment was done by, among others, Karl Lashley of Yerkes Laboratories of Primate Biology, who was famously searching for the physical location of memories. Subsequent workers have pointed out limitations of Lashley's studies, but it turns out that he was headed in the right direction.[1]

During the same era, clinicians made a second type of observation, this time on human patients. These clinicians reported two kinds of developmentally induced blindness. In the first, a patient who from birth had had one eye occluded—from a cataract, for example, or from rare eyelid problems—but then had that anatomical problem removed still ended up with one blind or nearly blind eye. Something about the early occlusion kept that eye and its central neural pathways from hooking up properly.

The second type of developmentally induced blindness concerned children who were born cross-eyed, with their two eyes pointing in different directions. When the children grew up, it was all too often found that one eye or the other had taken over: one eye worked, and the other one did not. This is called, colloquially, "sleepy eye" or "lazy eye"—the technical term is amblyopia. The eye is not truly blind—you can show by specialized tests that the retina is working—but the person has no useful vision through that eye. (There are various therapies for this condition at present, the commonest of which is to patch alternating eyes during early childhood so that one eye never has a chance to take over and suppress the other.)

The pioneers David Hubel and Torsten Wiesel, mentioned as discoverers of image processing in the visual cortex, undertook to repeat these experiments in animals and discovered the neural basis of the lazy eye: during a critical period in early life, the synapses that connect the retinal output to the central nervous system are malleable. If cortical neurons get a lot of conversation from one eye and none from the other eye, axons representing the first eye grab all the attention of the cortical neurons—their synaptic space—leaving the second eye functional but with no cortical neurons to talk to.

For crossed eyes, they found, it's a bit more subtle. In normal circumstances, images from one eye and images from the other eye are almost perfectly in register, and the same spot in the visual scene stimulates a single group of cortical neurons. When Hubel and Wiesel artificially crossed the eyes of animals, however, making a young animal wear a prism that shifts its visual image, images from its two eyes did not properly converge on the same brain target. If the eyes deviate so that they do not look in perfect parallel at the world, the central input is shifted and no longer creates a single cortical map. The person sees double, literally. They see two separate images. In the case of crossed eyes, the brain is faced with a problem: the images from the two eyes conflict. The brain has to choose one eye or the other. Connections from one eye are suppressed—first temporarily but after a while permanently, leaving that eye functionally blind.

A clever experiment demonstrates a different kind of reorganization of cortical responses. Under normal circumstances, there is a "map" of the retina on the visual cortex. To be sure, it is distorted by the undulations of the cortex's surface, but you can see very directly that neighboring points on the retina project to neighboring points on the visual cortex, creating an organized map of the visual scene on it. The experiment was to painlessly make a very small hole in the retina of a monkey using a laser. The experimenter, Charles Gilbert of Rockefeller University, then recorded from the visual cortex, to see how the cortical map had responded. Initially, there was a hole in the cortical map, corresponding to the hole in the retina. After a while, though, neighboring regions of cortex moved over to occupy the vacated cortical space: neighboring regions of the retina communicated with the cortical cells that normally would have responded to the damaged region.

This does not mean that vision was restored for the damaged region of retina. If you have a lesion in your retina, you are never going to see anything in the region that was destroyed—you have a blind spot. But even though the brain can never compensate for the hole in the retina, the region around the retinal lesion will "own" more cortical neurons than it did previously, and this predicts that it will do its work better. To my knowledge, this prediction has not been tested; at the very least, though, the hyperinnervated region should be more robust—and perhaps less sensitive to further damage than before.

One way to think of this is as nature's way to prevent cortical idleness. If an area of cortex is no longer receiving inputs from its natural place, it would be wasteful for that area of cortex to be forever inactive. Instead, after a while, its function is given over to undamaged inputs. In the more general case, you can easily imagine this mechanism as a way of dealing with small strokes. (Neuropathologists tell us that we all incur these small losses of brain tissue during the course of our lives.) Imagine that you have a tiny cortical stroke, affecting only a very small blood vessel, and that the region of brain it feeds dies. It would be wasteful of precious cortical resources for areas of the brain that used to receive input from the region that is now damaged by the stroke to be forever silent. Instead, the brain makes the best of a bad situation by giving those brain areas over to their neighbors.

REORGANIZING NORMAL PERCEPTION

In the previous section, we saw how the senses adapt to various types of neural damage, which are pretty crude events on the big scale of neural life. But there are subtler reorganizations that occur naturally and happen to all of us.

We learned a lot when experimenters got access to brain scanning. One of the striking indications of brain plasticity came from the brain activity of people who had been blind from birth. When blind volunteers used their fingers to read Braille while in the scanner, the brain areas usually occupied by processing visual input—again, the primary visual cortex—were activated. These people had used their touch intensively for years and years. Somehow, the processing of tactile information had taken over the unused visual center.

Another dramatic example, this time with sighted people, came from a study of violinists. To play the violin, you make large, relatively crude motions with one arm as the bow sweeps up and down across the strings. With the other hand you make a series of very subtle movements, depressing the strings at varying, tightly defined locations up and down the violin's fingerboard—very quickly if you are a good violinist, astonishingly quickly if you're a star. This is a remarkable task for the speed and precision it requires. Professional violinists practice these movements for hours each day. This has a consequence on the physical arrangement of the connections in their brains, because as you may have guessed, movements of the fingers are controlled by a specific brain area. In professional violinists the area concerned expands, even pushing aside functions from neighboring brain tissue. But this only occurs for the hand used for fingering the strings. The same regions on the other side of the brain, which controls the other hand, have no expansion, because the required movements of that hand are relatively crude, even in professional players. The regions of the brain that control finger movement on the arm-moving side remain entirely normal.

(Violinists are an extreme example, but I wonder what happens in other cases, too. If you are a professional athlete, do your muscle-control brain circuits expand at the expense of others? If you spend much of your working life worrying about the brain, do the worrying-about-the-brain circuits expand at the expense of the appreciating-opera circuits?)

The opposite situation—deprivation rather than overuse—occurs in the laboratory. There, cats raised in darkness lost the ability to properly fuse images from their two eyes. Slightly more controversial were experiments in which cats were raised under conditions where the only patterned vision they were allowed was of a pattern of vertical stripes or horizontal stripes. Orientation selectivity is weak at birth under any conditions and sharpens during postnatal life. In the extreme case, stripe-reared animals grew up with a bias in the orientation selectivity of the neurons in their primary visual cortex: an abnormally high number of cells were tuned to specific orientations, vertical if the cat's only visual experience had been vertical stripes, horizontal if the cat saw only horizontal stripes.[2]

A clever variation on dark-rearing was to deprive animals during early life of the ability to see motion. The experimenters did this by rearing cats in an environment lit only by very brief strobe flashes. This allowed the cats to see the usual world, but the flashes were too short for any meaningful movement of objects across the retina to occur—in other words, these animals' cortices were deprived of visual movement. What happened? These animals grew up without direction-selective neurons in their cortex.

A final and critical set of experimental manipulations confirmed directly the role of synaptic plasticity in visual development. These experiments were done by Michael Stryker, Carla Shatz, and their students, now at UCSF and Stanford, respectively, and involved the lateral geniculate nucleus.

One section of the LGN is dedicated to input from one eye, and the other to the other eye. However, our LGN does not start out that way when we're born. In normal infants, axons from the two eyes branch widely and each covers a wide spread of the LGN. There's little segregation into right and left at first. Segregation is created by the patterns of activity of axons arriving from the eye. Even before birth, these axons are firing—in bursts, like an automobile engine on an extremely slow idle. These bursts are synchronized: activity from one eye arrives at the LGN at one time, and activity from the other eye arrives at a different time.

This sets up a situation for a critical form of plasticity—the antecedent to machine learning. As you will learn in Chapter 9, Donald Hebb had proposed that groups of neurons that fire together have their connections strengthened. When lots of retinal axons from one eye are driving their LGN target cells at once, the synapses between those axons and the LGN cells are strengthened relative to synapses from the other eye. Gradually, the initially wide-spreading axons refine their LGN targets, so one clump of LGN neurons becomes responsive to inputs from the right eye, another to input from the left eye. The conclusion is that the map of the retina on the LGN is sharpened just as postulated, by strengthening of concurrently activated synapses. In confirmation, when Stryker used a drug to block the activity coming from one eye, this sharpening never happened.

All of these findings point to malleability in the organization of the sensory systems. But how important is this under natural human conditions? At the risk of pounding this point into the ground, what happens if a person grows up without any vision?

LEARNING TO SEE

Donald Hebb predicted that vision is to a major extent learned: complex perceptions are formed from experience, by association, because objects in the world occur in clusters of individual features. He believed that this had to happen early in life or, as some evidence at the time suggested, the brain afterward would become unable to form the necessary new assemblies. His basic idea was right: much of vision does depend on visual experience. But his conclusion that this had to happen at a young age seems to be only partly true.

The evidence comes from experiments in which individuals blind from birth were later given sight. Pawan Sinha of MIT, a native-born Indian, realized during a visit home that there were perhaps three hundred thousand children born with dense congenital cataracts in the villages of India. In these children, the lens of the eye is replaced by a cloudy fibrous tissue. The cataracts allow in light and dark but deprive the child of all detailed vision. In a brilliant combination of humanitarianism and science, Sinha organized a program to search for these children and transport them to New Delhi, where surgeons in a modern hospital replaced their lenses with clear synthetic ones—the same cataract operation carried out for many aging individuals in the developed world.

Sinha's team tested his patients' vision before the operation, immediately after it, and months or years later. Taking away the cataract did not immediately restore detailed vision in the children. The world to them seemed a confused blur. But as time passed they began to see, and after a few months they could see details beyond simply light and dark. Many could walk without a white cane, ride a bicycle in a crowded street, recognize friends and family, attend school, and carry out the activities of a sighted person.

Yet their vision seems never to have become perfect. Their visual acuity remained below normal, even after months of training. One commented that he could read headlines in the newspaper but not the finest print. Some had trouble with specific visual tasks, such as separating two forms that overlap each other, as in the image below.

Most of us see this image as a triangle partially overlapping a square, but some newly sighted people only see the lines as a single, complex object. (Interestingly, the problem is "cured" if either the triangle or the square is caused to move 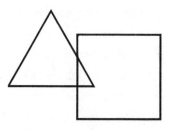 independently of the other. This and other forms of visual training seem to help the recovery of vision.)

So it seems that much vision can be restored, but several findings remind us that the plasticity of the visual system is not limitless. First, the fact that the face patches have reproducible locations in different individual people (or monkeys) shows that the brain has some level of intrinsic pattern for them. Second, as the newly sighted Indian children learned to see, their brain patterns underwent a change. Right after the cataract removal, fMRIs showed a disorganized, widely spread response to visual input, including faces, but it quickly changed to a series of patches—and the patches were in their normal locations. Again, this shows that the brain knew ahead of time where the face patches were supposed to be; it is evidence for at least a low level of predetermination of the visual structures. Livingstone calls these predetermined locations "proto–face patches."

Finally, a powerful and elegant experiment on sensory neural plasticity was published in late 2017 by Livingstone and her colleagues. They raised monkeys from birth in an environment where they never got to see a face. Not a human face, not a monkey face, no faces at all. To accomplish this experiment sounds like a lot of work and probably was, but in principle it's not too hard. The monkeys were cared for, lovingly, by the experimenters. Whenever they were near the monkey, however, the experimenter wore a welder's mask—a great curved sheet of darkly tinted glass that reaches from the forehead to below the chin.

The monkeys otherwise grew up in a completely normal visual world: they could see everything in their cage and in the surrounding room; they could see the experimenters' torso, arms, and feet; they could see the baby bottle with which they were fed. They could hear the normal sounds of a monkey colony. Their only deprivation was that they never saw faces. These monkeys developed in most ways normally, and when they were introduced into the monkey colony after the experiment was finished, they socialized happily with their peers and integrated successfully into monkey society.

After the experimenters trained these monkeys to lie still inside the fMRI scanner, they then tested the monkeys by showing them various things, including faces. As you may have guessed, they grew up without face patches in their brains. Remarkably, though, what would normally have been the temporal lobe's face recognition areas instead responded to images of hands. In a normal social environment, the most important visual objects for a primate are faces. Faces signal anger, fear, hostility, love, and all of the emotional information important to survival and thriving. Apparently, the second most important feature in the environment is hands—the monkeys' own hands, and the hands of the experimenters who nurtured and fed them.

Although what would normally have been face patches turned into "hand patches," this preference was still somewhat plastic. About six months after the monkeys were allowed to see faces of the experimenters and of other monkeys, the cells in the face patches gradually reverted to being face-sensitive. Evidently, faces convey so much important information that they recaptured the brain territory that had been taken over by hands.

The existence of face patches explains a curious and long-recognized clinical observation. There is a condition known as face blindness (prosopagnosia, from the Greek *prosop*, "face," and *agnosia*, "ignorance") in which a person's vision is quite normal except for difficulty recognizing faces. The sufferer can see fine, is as good as anyone else at distinguishing one face from another, but has difficulty recognizing faces from memory.

There are gradations of prosopagnosia ranging from almost complete, which may bring the person to medical attention, to very mild. At the opposite pole are people who are super face recognizers. One of his aides commented that Senator Edward Kennedy could recognize ten thousand people. Speaking personally, I am well on the prosopagnostic side. It is an embarrassing problem. I can spend a pleasant evening at dinner with you, and the next day pass you in the hall and think, "Do I know that person?" but can't get any further than that. So if I have cut you cold at one time or another, please understand that it was my disability speaking, not any lack of interest in you.

To recap: face patches are distributed widely, seem to function in concert, and come to be face-sensitive through experience. As Livingstone points out, this widely distributed, experience-dependent system behaves in many ways as though the cells were involved in some sort of learned neural circuitry.

9 | Inventing the Nerve Net: Neurons That Fire Together Wire Together

> My problem was to understand . . . how it was
> that a patient could have a large chunk of brain
> tissue removed without much effect on his IQ or his
> intelligence as it seemed to his family. How could
> there be no loss of intelligence after . . . removal
> of the entire right half of the cortex?
> —D. O. HEBB

AT THIS POINT WE'RE GOING TO START TRYING TO PUT THINGS together—to give whatever answer we can to the question of how you recognize your daughter. To get to the basics of it, we'll jump back in time, to the 1960s, and show the history of an exciting branch of neuroscience and computational biology. The basics

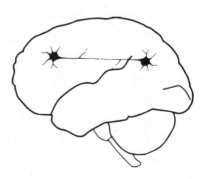

of it were imagined more than a half century ago by a Canadian neuroscientist named Donald Hebb.

Imagine a brain that contains only two neurons. They are connected, as most neurons are, by a synapse. (Let's not worry for now about how this brain talks to the outside world. It is a hypothetical brain.) Imagine that the two neurons are activated. For now, "activated" just means they fire nerve impulses.

Now let's compare two different situations. In the first, the two hypothetical neurons are activated, but there is no relationship between when neuron one is activated and when neuron two is activated. Each fires along in response to inputs without regard to the other. In the second case, whenever neuron one fires, neuron two fires as well, and vice versa. In both cases, the two neurons show the same amount of activity. The only difference is that either they fire at the same time or they fire at unrelated times.

As it turns out, when the two neurons fire asynchronously, nothing special happens. But when they fire together, something dramatic occurs: the synaptic connection between the two is strengthened, so activity in one of the neurons becomes able to trigger activity in the other. To make the rule easy to remember, someone phrased it as "Neurons that fire together wire together."

This is a simple little piece of neuroscience. I've just described it to you in under two hundred words. As it happens, though, this coincidence-based strengthening of synaptic connections is the most basic form of memory, from which is built all perception, emotion, and action. Donald Hebb was one of the founders of modern neuroscience. He described his ideas about nerve nets in his book *The Organization of Behavior*, published with extraordinary prescience in 1949. The modifiable connection between two neurons has come to be called the Hebb synapse. We'll circle back later to its descendants in modern nerve nets. A Hebb synapse—or some silicon version of it—powers most of the modern forms of artificial intelligence.

HEBB'S ORIGINAL NERVE NET

In *The Organization of Behavior*, Donald Hebb proposed a very broad theory, encompassing many facets of neuroscience—a theory that laid the groundwork for what is now known as machine learning. It is the key to how animals survive in their natural world. But his first love, the bedrock of his thinking, concerned perception.

Hebb started with simple kinds of perception. Take, for example, a simple line drawing—let's say a black outline square on a white background. We see it as a unit, a continuous figure, but in reality, the outline of the square, viewed on a fine enough scale, consists of a row of dots. Our retinas detect the black line as a row of dots because the black line falls upon an array of individual light-sensitive neurons, the retina's rod and cone photoreceptor cells. They are closely packed in a regular array on your retina. No matter how smoothly the line is drawn, the retina detects it as a row of dots.

But that is of course not what we see. We see a black square on a white background, because our brain joins the dots into lines. Furthermore, we do not perceive each of the four lines that join to form a square individually; we see the square as a unit. Only in a damaged brain can the unity of our perceptions fall apart: MIT neuropsychologist Hans-Lukas Teuber reported on one patient who saw, instead of a passing motorcycle, a string of motorcycles. In a normal brain, things tend to be seen as wholes rather than as assemblies of parts. But the world does not intrinsically tell us what is an object; it only shows us an array of pixels, and the brain has to figure out what is an object. In Teuber's example, the brain figures out that it is seeing a moving motorcycle rather than a string of stationary ones. This principle, called closure, led Hebb a step closer to his understanding of memory.

Psychologists of the early twentieth century knew that objects had a gestalt, an identity that supersedes variations in individual details. A favorite example is that leaving out a piece of a familiar image does not destroy the integrity of the image. A triangle is instantly seen as a triangle, even if there is a gap in one of its sides.

Hebb and his students were the first to describe an intriguing manifestation of this phenomenon, using what are termed stabilized retinal images. Even though we are not aware of them, the human eye continuously makes very small movements—a sort of tremor of the eye. These are different from the eye movements that we make when we look from place to place; they are so tiny we do not even notice them. But these tiny movements affect our visual system, and our vision. To stabilize a retinal image, Hebb and his team used a tricky contact lens constructed with a tiny, tiny stalk sticking up from its surface. An even tinier lens at the end of the stalk focused test images on the retina. This contact lens eliminated the effect of small eye movements on the image as it usually hits the retina.

Most neurons of the retina and visual portions of the brain are not very interested in objects that are unchanging. They respond well when a new object is shown, but gradually stop responding if the image shows no change at all. This is useful; it means the brain doesn't spend energy on things that carry no new information. But a side effect is that images are prone to fade from perception if they don't move. The resting tremor of the eye counteracts that process by making the image slip back and forth across the retina (again, in movements too small for us to perceive), so the neurons do not fatigue and the object is continuously visible. But the experimental contact lens nullified those small eye movements so that whenever the eye moved, the image moved as well.

As you might have guessed, when the experimental contact lens was used, the image disappeared because there was no tremor to stabilize it. But to Hebb, the critical result was not that it disappeared. What was most important was what happened while it disappeared. The stabilized image did not disappear in an unorganized way—for example, decomposing into a scatter of dots. Instead, it disappeared in chunks. A whole square might disappear at once. Alternatively, an outline square might initially lose one of its four sides but keep the other three, then lose two more sides, and finally lose the last remaining line.

Hebb postulated that these chunks correspond to the concurrent activation of groups of neurons in the brain. These he called "cell assemblies," his version of the most basic nerve net. Later we will get to Hebb's ingenious idea about how cell assemblies are created, which turns out to involve the celebrated Hebb synapse.

Hebb's postulate of a cell assembly actually served two purposes. First, it explained why perception occurs in chunks. In his theory, the brain's image of a square consisted of activation of cell assemblies for each of the four sides, linked together synaptically. But the cell assemblies that join together to form a square do not have to be located in a physical square within the brain. (In fact, cell assemblies, and their higher-order linkages with other cell assemblies, are widely distributed throughout the brain.) Hebb worked out a scheme in which very simple elements, like our square, could be represented by very simple cell assemblies, but the cell assemblies could be linked together to form far more complex objects, or even to form thoughts and conscious memories. Our specific square could originate as the sensory representation of four lines, but by being linked to other cell assemblies, it becomes part of a memory of the square.

In this book we are going to restrict ourselves to exploring simple perceptions, but a point worth noting is that Hebb's cell assemblies are linked not only to other events of the same type (i.e., vision) but also to cell assemblies that might reflect sounds, smells, and tastes. Emotions, too—the whole web of experiences that make up each person's history. Cell assemblies can be linked to distant memories, and of course they contain not a few but hundreds of thousands, or even millions, of interconnected neurons, located near and far in the brain. These include conceptual and emotional associations. If you are a fan of Proust, you'll understand how the taste of a madeleine led him to a whole memory of lost times.

This spiderweb-like associative property is the first thing many of us think of when we hear the term "nerve net." And when it's viewed from a distance, the brain's web of connections would indeed look that way. But within that web are highly specific connections, and those are the ones with implications for the fundamentals of vision.

Hebb's idea had a second and fascinating implication, quite separate from perception. It explained the amazing resilience of the brain in the face of damage. In 1937, the Montreal surgeon Wilder Penfield was carrying out pioneering procedures in which pieces of the brain containing abnormally firing neurons were removed so as to quell epileptic seizures spreading from a damaged area to more normal parts of the brain. This is an effective therapy that often does control the patient's seizures. But if surgical removal includes specific sensory or motor areas, the patient also loses something important: vision or hearing or touch, or the ability to walk or carry out skilled movements with part of the body. For this reason, Penfield carefully mapped the brain during surgery, with patients remaining conscious throughout the procedure, to make sure these specific sensory and motor areas were avoided.

This left him most of the surface of the brain to work in: large areas that are not part of our essential sensory or motor systems. The functions of these other areas were, and still are, poorly understood. Penfield hired Donald Hebb to find out if there were hidden deficits caused by damage to these "silent" areas. Did removal of any particular diseased segment have consequences for intelligence? Penfield knew that if there were losses caused by removals of tissue in these unexplored areas, they would be subtle, because when he spoke with the patients after the surgery he found their communication remarkably little impaired.

Still, no sane person would imagine that large areas of the brain are useless. Penfield wanted Hebb to find out what had been lost, even if it was hard to detect through everyday conversation. Subsequent decades of research have finally shown that there are indeed subtle losses of function following most brain lesions. But what impressed Hebb so deeply was what had *not* been lost: brain lesions did not seem to cause the loss of specific memories.

To avoid confusion, it is important to distinguish between the loss of specific memories and the loss of particular abilities. Most of us know, from our reading or from tragedies affecting our friends and families, that damage to the brain can cause very specific disabilities. The ability to understand language may be lost (aphasia); one limb may be numbed or paralyzed; the muscles in the face may droop. The loss can be remarkably specific: one's ability to speak may be lost, yet one might completely retain the ability to understand spoken language. But these are not the same as the loss of individual memories, which is what Hebb was studying. They are losses of abilities.

As Hebb memorably put it, "We do not lose the memory of the red swing." If your grandmother's house had a porch, and on the porch was a red swing, you might lose the whole memory of the porch (or the farmhouse, or the farm). But what does not seem to happen is that you retain a bright image of the porch with the red swing gone.

More generally, Hebb observed, patients may lose a lot of old memories (or they may lose the ability to form new memories), but when that happens, all the parts of those memories are lost together. There can be a kind of "graying-out" of memories (if this never happens to you, I congratulate you), but for every memory we do retain, we retain most of its elements, its particular pieces.

This seems to imply that memories do not have a place in the brain. But how can this be? Memories are not ephemera—spiritual things that float outside your physical body. They most certainly do have to live somewhere, and that somewhere is clearly in the brain. The computer we call the brain contains the memories, but specific individual memories do not seem to have a physical place in the brain where each is stored.

Hebb's notion of nerve nets not only accounts for the unitary nature of perceptions but also, *at the same time*, explains why Penfield did not find bigger deficits after removing pieces of his patients' brains. Let me explain why.

Nerve nets are distributed through many regions of the brain. If a nerve net is not tightly localized but widely spread out, it could explain why individual memories don't have a fixed location. If a cell assembly consists of a very large number of neurons, all interconnected, losing a few would not hurt us much. Most of the cell assembly would remain, still interconnected, and would continue to represent a perception, or a memory, or a thought. That has to be why memories do not live in a particular place: they are distributed throughout the brain. To answer Penfield's question, damage to a nerve net does cause a decrease in function, but it is so small as to be unmeasurable (especially in Penfield's era). Because of nerve nets, the brain becomes what engineers call "fault tolerant."

TEACHABLE SYNAPSES CREATE THE CELL ASSEMBLY

So far I've told you how cell assemblies create perceptual wholes from parts, and how a nerve net distributed widely around the brain explains the slippery location of memories. At this point, this is just a cute theory. What elevates it to immortal—the reason Hebb's theory is remembered—is his idea for how a cell assembly might get built: by the simple and elegant action of the Hebb synapse.

Suppose we have a cell assembly widely distributed around the brain, and its job is to represent and store a perception of an object. A cell assembly is a miniature nerve net. How does this particular nerve net come to correspond to the brain's perception of that particular object in the outside world? In other words, what is the relationship between the cell assembly in the brain that represents the four sides of a square—the square that does, in fact, appear in the world as a unit—and the experience of looking at squares?

Because the world is orderly, the brain gets its inputs in an ordered way. The world is projected on the visual cortex of the brain topographically—there is a map of the visual world on the brain's surface. More precisely, there is a map of the retina on the surface of the visual cortex. Thus, a line activates neurons arranged roughly in a line on the visual cortex. The visual cortex is the main portal through which information passes to object-recognizing centers. In consequence, when a line is viewed, neighboring cortical neurons fire together. You can guess the second part: because this group of simultaneously firing neurons are connected by synapses, those synapses are strengthened by their firing together, and they "wire" together. They form a cell assembly representing a line. Once they've done this, whenever one or a few of the connected neurons fire, the rest will tend to fire. When the cell assembly started out, all of its synapses had the same strength; after repeated stimulation by the line, a subset of them—those corresponding to the line—are strengthened.

Once this happens, the cortex becomes biased to see lines. If it sees part of a line, the now-connected neurons tend to go off together and report to the rest of the brain that the stimulus is a line. The cortex's report is likely to be true because lines occur much more frequently than by chance in the world. A truly nonredundant scene—without any skewing of the frequencies—looks like the snow on a messed-up television screen. Regularities in the world skew the brain's interpretation of inputs in favor of those regularities.

At the next level of complexity, four lines occurring in a conjoined fashion (a square) also cause correlated firing—in this case firing representing a square. At this point, my simple two-dimensional illustration starts to break down because we have to get beyond any particular square to a cell assembly representing a gestalt (an identity that supersedes variations in individual details) for "squareness," and that cannot be represented in a simple way. We will address this problem later on, so for the moment let us leave it at this: cell assemblies for simple geometric forms can be concatenated into a larger nerve net that abstracts "squareness" from particular instances. Hebb thought that this kind of perceptual learning (which is quite unconscious) was the foundation of all perception.

———

HEBB'S EMPHASIS ON lines turned out to be unfortunate for neuroscience's intellectual history, because of evidence in the 1960s—not very good evidence, as it turned out—for innate line-detecting neurons in the brain. This seemed to put paid to his entire theory. But Hebb's basic principle—the Hebb synapse and how it can build the brain's connections—did not die. It slumbered for a decade or so, and then made a spectacular comeback, with a revival of interest in nerve nets.

DONALD HEBB IN MONTREAL

In 1965, Donald Olding Hebb—addressed by colleagues and students alike as "D.O."—was a man of medium height and slight build. A serious bout of bone tuberculosis had left one leg shorter than the other, so he limped. His posture was erect and he held himself firmly, yet the asymmetry of his legs gave his silhouette a curve, like a parenthesis. He had sandy hair that was, at sixty, turning gray, and fair Scots-Irish skin in a square Scots-Irish face.[1]

Hebb was born and raised in Chester, Nova Scotia, a sparse village of simple frame houses on a rocky harbor of Nova Scotia's south coast. The first Hebb (then spelled "Heb") had emigrated to densely forested Nova Scotia from Germany in 1753. His many descendants spread throughout the province, and there is a town called Hebbville not far from Chester. These Germans intermarried with the more numerous Scots, and D.O. proudly considered himself of Scots ancestry.

Donald Hebb was born in 1904. His father was a town doctor. Hebb's mother, the former Mary Olding, was also a physician, having matriculated at Dalhousie University in 1892. She was one of the first women to graduate in medicine in North America, and it is more than likely that some of Hebb's independence of spirit came from her. His sister, Catherine, became a renowned neurochemist and spent most of her career at the University of Cambridge in the United Kingdom.

D.O.'s family lived in a plain, rectangular house on Chester's main street, a few steps from the harbor. In later years, various Hebbs would return to vacation in Chester; the Hebb family still maintains a house nearby. One of D.O.'s passions was sailing small boats. A current Chester resident recalls that her brother had a job maintaining the Hebbs' boats. He came home complaining that the Hebbs had "very high standards."

Hebb did have high standards, but what he cared most about was originality, and one of his overriding messages was to shun the pedestrian. He was looking for ideas that were truly new, and his department made remarkable discoveries: "pleasure centers" in the brain; brain regions that cause selective loss of recent memory; brain neurons that tell a rat where it sits in space; the "kindling" of epileptic seizures. Every one of these was an out-of-left-field discovery. Not one was a merely incremental improvement on what had come before. My McGill classmate John O'Keefe won a Nobel for daring to record from a rat while it ran a maze, figuring out how to do the recording (in 1966, no less) and noticing the strange behavior of the "place cells," which identify the animal's location in space.

A second message Hebb pounded home to everyone was the importance of scientific presentation. He cared about writing—his original ambition in life had been to become a novelist. By the time I knew him, at least, he no longer wrote fiction—he commented jokingly that his great work of fiction was *The Organization of Behavior*. But among his recreational reading was pulp science fiction, which he devoured in huge gulps. Every few weeks, a box of Hebb-read sci-fi paperbacks would appear outside his door for the taking.

"Listen," he said, "presentation is part of doing science. Your private knowledge is no good to you or to anyone else. You have to make it available. It doesn't matter how good your study is or how clever your ideas are; if nobody reads about them, they are a waste. You need to be persuasive." I imagined Hebb crafting *The Organization of Behavior*, all the while thinking, "If I do not sell these ideas well, they will go nowhere." And he did sell them well: part of the charm of that book, which was an immediate success when published, is the lucidity of Hebb's prose, its clarity, and its subliminal good cheer.

Under his formal and disciplined exterior, Hebb was a gentle man. His secretary at that time was a strikingly tall redheaded young woman who was said to have been a jazz club bartender before coming to work for Hebb. I can't speak to her history, but it would have been Hebb's delight to hire a bright young woman who flew in the face of convention. (Certainly she was well acquainted with the excellent jazz and blues scene of downtown Montreal.)

Hebb was extraordinarily creative; even his messages about creativity were creative, which is also to say that they were unusual. He firmly believed that learning unnecessary things was bad for you, which he buttressed with the old joke about the aging professor of ichthyology who reported that every time he learned the name of a student, he forgot the name of a fish. This belief was probably based on principle: since Hebb thought memories were formed by modifications of synapses, sooner or later you would run out of synapses.

A corollary was that graduate students should not be forced to take formal courses. (My biggest reason for choosing McGill for graduate school was the chance to study with Hebb; the second most important reason was the lack of required courses.) The only inflexible requirement for the PhD in his department was a light course in statistics, which was demanded by some higher credentialing body. Beyond that, there were no requirements at all.

PERCEPTUAL LEARNING

In the 1960s, what Hebb really cared about was perceptual learning. Now, the phrase "perceptual learning" can mean a lot of things, and I want to take care to distinguish them. One interpretation is commonsensical: we get better at performing a sensory task the more we do it. A classic example is the ability to identify the location of a pinprick on the skin. This is tested by lightly touching the skin with two pins at varying distances from each other. The closest spacing that no longer lets you distinguish two pinpricks from a single one is the two-point discrimination threshold. If you let someone test you on this task every day for a week, you will find at the end of the week that you can distinguish pinpricks more closely spaced than at the beginning.

My favorite example of perceptual learning in everyday life is the way we can learn to hear music. I have listened to a lot of music over my lifetime; I now hear music far more clearly than when I was younger. This is true not only for the voices of a Bach partita but also for the lyrics (unfortunately) of a heavy metal rant. I do not care at all about pop lyrics, nor for that matter do I have a deep understanding of the formal construction of Bach's music. Hearing these things better is not a matter of understanding; it shows that our sensory experience itself can be "educated."

Obviously, there's a simple explanation: you may just pay closer attention to your skin or to Bach after you practice. But there is a subtler, deeper meaning of perceptual learning, and this is the one that interested Hebb: *the neurons of your sensory system adjust themselves to the stimulus.* In fact, Hebb believed—in 1949, before the advent of information theory—that neural reorganization in response to the regularities of the natural world was not just icing on the cake but a fundamental basis of perception itself.

We have already encountered regularities in the natural scene when we looked at how we learn to see a line. The world contains many visual events that occur redundantly. Look out your window: on a clear day, the blue sky looks about the same from point to point. That is what is meant technically by "redundancy." It is unchanging from one point to another, and thus the visual signal is the same. The same is true for other, more complicated, things. An area on the fender of a red car is likely to be surrounded by something else red (in this case, neighboring places on the same fender). This, too, is redundancy in the sense we're exploring.

To get a bit fancier, think about a forest, or a telephone pole, or a building. An important feature of all three is that they contain straight lines. A straight line is redundant: if you pick one point on the line, the neighboring point is more likely than chance to be along the same line. On the other hand, if you pick a point that deviates from the line, the signal will be different. So a line is, compared to the clear blue sky, only partially redundant—the position of a point is restricted in one dimension instead of two—but the principle is the same. In fact, there are very few visual inputs in our world that are not redundant. A truly nonredundant scene, again, looks like snow on a television screen.

In terms of the demands that vision puts on your nervous system, this is a very big deal. It means that your perception does not really need to evaluate every pixel in a scene. It can use the regularities of the scene to predict what neighboring pixels contain, and insofar as the brain knows those regularities ahead of time (because it has created cell assemblies representing them), it can save itself work. This is a huge general principle in the design of biological sensory systems. We saw this when it came to edges, and we'll see it again when it comes to viewing complex objects like faces.

Thinking about these regularities is what led Hebb to postulate they were mirrored in a cell assembly. As we have seen, cell assemblies are groups of neurons that fire together because of exposure to regularities in the world. They are sets of neurons that have been excited together, so neighboring neurons come to fire together. If there are four activated pixels in a row, the cortical neurons responding to them will become linked. In effect, the brain predicts that the first three pixels will be accompanied by a fourth one located on the same line. Of course, the prediction may be right or wrong: the line may swerve, and so the fourth pixel is, in fact, off to the side. But the likelihood is weighted toward the predicted event. Because the visual world is primarily composed of regularities, your brain's guess is going to be right more often than not, and the efficiency of your perceptual guess is improved.

Another way to say this is that all perception is biased by things you have seen before. I mean here not just that you expect to see certain things in certain contexts—the way you expect to see your cousins on Thanksgiving—but that at an unconscious, basic level, even the simplest elements of sensation are *created*, just as much as they are literally recorded, by your eye, ear, skin, or nose.

As it happens, an experiment that I did at McGill describes a primitive form of this "creation." I was studying what is called a motion aftereffect. These happen when you look at a moving pattern fixedly for a long time and the movement then stops. It looks like the pattern begins drifting in the opposite direction. (An alternative name is the "waterfall illusion": after you look at a waterfall for a while, anything you look at afterward appears to stream upward.)

This kind of perceptual illusion has been known for a long time. The new information my experiment added was that the motion aftereffect could last, remarkably, for days. That meant that it wasn't just some sort of fatigue, like the short-lived aftereffects of looking at a bright light for a while, but a more enduring bias to perception. I also discovered that it was linked irrevocably to the original stimulus. In everyday life, perception remained quite normal; only when you looked at your original moving pattern (now stationary) did you see the creeping opposite movement.

Still more remarkably, I found that for this to happen, the stimulus had to fall in the same part of your gaze as before. Because the brain's visual centers are organized like a map, long-term change in vision happens not in any vague world of expectations or consciousness or what have you, but in fairly simple parts of the brain's sensory systems. The subjects of these experiments have no conscious hint that their perceptions have been changed—their vision is entirely normal—until they are shown the strange movement of the test stimulus. For that particular object, their perception has been changed—"created," if you will—in a highly specific and more or less permanent way, just as Hebb predicted.

The phenomenon that I had discovered was slightly kinky. The way the test stimulus moves is otherworldly. It looks like it's moving, but you know it really isn't because you can compare its position with surrounding stationary objects. It made for an amusing parlor trick. Later, however, similar results were demonstrated by Charles Gilbert and his colleagues at Rockefeller University for perceptual tasks that measure the ability of a person to make a meaningful everyday discrimination—the position of a line in space. This shows that useful everyday experience can be changed by experience.

PROOF OF THE CENTRAL POSTULATE

Ironically, Hebb's theories of perception and cognition are less famous than his postulate about synaptic plasticity, the Hebb synapse. It's understandable that those theories fell out of the limelight: they're harder to grasp than his ideas about the details of an individual synapse, which is an actual thing with (microscopically) visible moving parts. Too, his theories of perception were untestable at the time. Only at the start of the twenty-first century has our goal of visualizing distributed neuronal systems seemed no longer a pipe dream.

Hebb did not worry much about the Hebb synapse. It was simply a postulate, a tool toward building his theory of perception and memory. In the twenty-first century, a synapse is an extremely specific thing. We know its geography in detail: a three-dimensional printer will make you a model of one you can put on your desk. We know the information that gets transmitted by a synapse, and we know the molecules that transmit it, of which there are dozens, each carrying out a specific task.

In 1949, Hebb didn't have any of that. The synapse had only recently become more than a postulate itself—something that was hoped to be true for its explanatory value, but lacking much direct evidence. Hebb could not do anything more with the synapse, so it was not worth getting bothered over. But he did propose that changes in synaptic strength might form the basis for memories.

And around 1970, it turned out that a change in synaptic strength could be observed in a brain subsystem simple enough for us to study with basic lab equipment. This change in synaptic strength is called long-term potentiation. (Scientists, when they are unsure of themselves, fall back on terminology that is strictly operational—"long-term potentiation" just describes exactly what happens in a certain type of experiment, nothing more. Nobody dared call it "synaptic learning," or the like.)

Long-term potentiation (LTP) directly demonstrates, using actual synapses, that repeated firing across a synapse makes the synapse grow in strength, for hours or even days after the stimulation has ended. This is just what Hebb had postulated: neurons that fire together wire together. Hebb's synapse, and theories related to it, have sparked a small industry of studies on the plasticity of individual synapses. One recent morning, a computer search for "long-term potentiation" yielded 13,800 published papers. In 2000, Eric Kandel won the Nobel Prize for his work on synaptic plasticity.

The first laboratory demonstration of LTP was by the Norwegian physiologist Terje Lomo. Lomo published a simple but accurate description in 1966. This was followed up in the early 1970s by a series of carefully detailed papers by Lomo and Timothy Bliss, as well as other collaborators. Lomo studied a brain pathway whose axonal fibers are easy to find for electrical stimulation; they can, through electrical stimulation, be made to excite a known set of postsynaptic neurons. (For what it's worth, the system is called the dentate region of the hippocampus, so named because its shape must have reminded some early anatomist of a tooth.)

The novel observation was that conditioning the presynaptic fibers ("upstream" of the synapse) with a high-frequency train of stimuli enhanced the postsynaptic cells' (downstream) response for a long period of time—hours or even days. This was extremely intriguing, because at that time the only neural events that had been observed took place on a timescale of milliseconds. The persistent changes that occur in long-term potentiation were an opening wedge, we hoped, to studying how memories work.

THE ECLIPSE OF THE NERVE NET

I hope I've conveyed how awed I am that Hebb figured out in 1949 what we only began to demonstrate empirically decades later. In 1949 nobody had ever thought of individual synapses as having memory. The idea of a "cell assembly" or "nerve net" of synapses was wildly far from the minds of early neuroscientists, who were preoccupied with figuring out simple reflexes—few-neuron pathways that transmit information in one direction only. More complex networks, though hinted at in the drawings of some early anatomists, had never been explored. And certainly nobody had sought to explain—in a single postulate—our recognition of objects *and* the endurance of individual memories when part of the brain was missing, as Hebb had done with his postulate of cell assemblies.

Furthermore, Hebb didn't have the advantage of the concepts we have from computer technology. Most educated folks now are at least passingly familiar with nerve nets as implemented in computers. These are the basis of much of machine learning, and they make it possible to find remarkable patterns in large data sets, which allows us to do things like predict epidemics before they happen, or model the evolution of stars.

Hebb did not have any of that. In 1949, brain scientists knew that damaging particular parts of the brain damaged particular abilities, but no one had dared to theorize in any detail about memory, or about the relationship between the brain and perception. Hebb leaped far into the future to ask questions we are only now beginning to explore.

As it turned out, he may have leaped too far into the future. For a couple of decades his postulate of nerve nets fell out of the limelight. But around 1960, it was taken up by the nascent field of computer theory, and several Hebb-like inventions were discussed. A particularly interesting one posited a computer called a perceptron, which we will discuss in detail in Chapter 10. But by that time Hebb's 1949 book on cell assemblies was old news, and Hebb did not formulate things in the quasi-mathematical way of computer science.

There were also several substantive obstacles to making use of Hebb's theories. First, it was hard to imagine the next steps following cell assemblies in understanding perception. How do cell assemblies become linked with other cell assemblies to create concepts and actions? How do complex perceptions (such as of a face) get assembled? Is Hebb right that there is a continuum from simple perceptions to larger ones—and are there cell assemblies that represent not just objects but thoughts? As I have said, sitting at your desk trying to understand things just by thinking about them is hard work. (It is much harder than laboratory experimentation, where the task is broken into small pieces.) Also Hebb, like most pioneers, pushed his ideas as far as he could. He had an intriguing notion about reverberating activity in cell assemblies, but it has had less staying power than the Hebb synapse. That and the extensions of the theory were a distraction.

The other big problem, and the one that eventually drove many neuroscientists, including me, to more directly biological approaches, is that we did not have any experimental tools that could deal with a distributed system, whose neurons are by definition firing all over the brain at once. We were thrilled even to be able to observe one brain neuron at a time. I have described that for you—a laborious process, yielding only a few significant observations of single cells per day.

There's no easy path from that to observing a nerve net, which by definition includes neurons widely spread around the brain. Together with many of my friends, I decided to leave cognitive neuroscience and retrain as a biologist in order to get down to the most rudimentary steps in perception. You have heard the results of our work; we now have a pretty good concept of the information that flows from the eye to the brain. Now let's see if nerve nets can help us understand what comes next.

10 | Machine Learning, Brains, and Seeing Computers

> Today every invention is received with a cry
> of triumph which soon turns into a cry of fear.
> —BERTOLT BRECHT

IN CASE I HAVE SAID THINGS TOO TIMIDLY, I, LIKE MOST SCIENTISTS, think that the brain is a computer. In early days, this was only a figure of speech. The computers of the 1980s could barely balance your checkbook. What happened next was that the power of computers doubled every eighteen months (as predicted by Moore's law). Computers of today take your breath away with their power and speed. Not only that; good, fast computers are relatively cheap. To be sure, the computers owned by the NSA, Google, and the CIA are bigger and faster than the ones you and I can buy. But we can buy computers fast enough to contain a brain, or at least a little piece of a brain. The result is that scientists can start to reproduce a brain, or at least to make a computer that can do some of the things that brains do—and these turn out to suggest specific rules that real, wet brains may be following.

This has been a wonderful development, with its own ironies. In the early days those slow old computers, trying their best to become artificial intelligences, were surely inspired by brains. But computers got better, and humans learned more and better ways to teach them—so much that they now outperform brains in some tasks. The consequence is that neuroscientists are learning a lot from computers about brains, as much as the other way around. It is a wonderful two-way street—forces joined in trying to understand perception—and at the end of the book we'll consider some ways in which the brain is still inspiring computer design. In fact, brains are pointing the way to the next big problem in AI, unsupervised learning. To get us grounded, I'll tell you first about some intelligent computers, and we'll then see what we can learn from them about the brain.

TWENTY YEARS AGO, I met at a cocktail party an engineer who worked for an aerospace company dedicated to computer vision. That company designed the systems that guide smart bombs. I asked him, "Do you ever use the principles of nerve nets in your designs?"

His scorn was apparent. "Look," he said, "my job is to make a system that can detect a tank in the woods. If my machine can't recognize the tank, I don't get paid. I don't have any use for fuzzy-minded bullshit about neurons. I have to be able to say how my machine works, end of story."

Vestiges of this way of thinking persist to the present, even when nerve nets are being used for tasks like steering a car—and even when some designers still can't say how their inventions do it. Engineers are practical par excellence—they want to see every step spelled out. Perhaps in the future, thinking computers will help them do so. After all, a computer is only a machine, with parts made of ordinary atoms: in principle it is always possible to find out what state it is in.

So I think they will figure it out eventually. In the meantime, though, smart computers, like brains, can accomplish tasks without our being able to say exactly how they did them.

COMPUTERS THAT LEARN

We have seen that the face recognition areas in a monkey's brain develop through experience: the monkey has to see faces in order to create neurons that selectively respond to faces. How does this happen? How does the repeated sight of a face translate into the actual hookup of neurons from the retina to the LGN, to V1, and onward to the temporal lobe? We have only glimpses and notions. We do have a firm biological basis for one aspect: long-term potentiation, the strengthening of synapses with practice. From another angle, we have an example we understand because we created it: computer vision.

The granddaddy of seeing computers (you might as well call them thinking computers) was something called a perceptron. Because perceptrons exemplify a basic principle of much machine learning, it will be worth our while to describe them here.

In a perceptron, invented in the late 1950s, a group of perceptual elements that can each detect only a single, simple thing provide input to a centralized, mechanized decision-maker. An early version of the perceptual elements got a folksy name from their inventor, Oliver Selfridge: "demons." Each demon shouts its little piece of information simultaneously; thus the name of his early version of a perceptron was "pandemonium." Each demon is responsible for seeing only a single feature of the total input. And each demon can pass along only three types of information: it can be silent, it can whisper "I see some," or it can shout "I see lots!" But that's all it can do.

The decision-making perceptron can get smarter if it is taught. Suppose it is learning to recognize a house. It is shown a picture of a house. All of the demons shout their piece of information. When the perceptron is told "That was a house," it reviews the input given by each demon. If a demon—say, one that recognizes straight lines—was shown the house and gave lots of output (because it saw lots of lines; houses contain lots of straight lines), the perceptron will give that demon's voice a bit more weight in the future. The straight-line demon cannot by itself identify a house, but the thing it can detect (straight lines) is correlated with the image of a house. Demons tuned to things unrelated to houses get their opinion underweighted, meaning that the perceptron pays less attention to their opinion in the future.

The picture shows the basic structure of a perceptron. In machine learning lingo, the perceptron is an algorithm for supervised learning of binary classifiers.

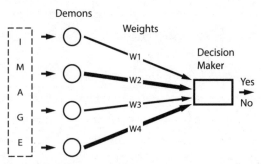

What that means is that the perceptron can make decisions like "Beth" and "not Beth." Although that sounds limited, you can imagine that running a bunch of them in parallel gets you some way down the path of object recognition.

A digitally formatted image is shown to the row of demons. Each of them has a particular feature to which is it sensitive. Each demon detects its feature, or not, and sends its message to the decision-maker, which collects all of the demons' outputs. A supervisor (teacher) then tells the perceptron that this was Beth or not Beth. At that point, by a process termed backpropagation, the perceptron adjusts the weights of the connections from the individual demons to the decision-maker:

demons that provided information useful for identifying Beth get their credibility increased. In the drawing, the thickness of the line weights is the equivalent of the credibility of the demons. (These are analogous to synaptic weights in a brain). The next time the "Beth" perceptron sees an image, the output of that particular demon to the decision-maker is overweighted; it is given more influence on the decision-maker. You repeat this many times, with many images of Beth, or whatever you are training it to recognize. Round and round the perceptron goes, getting more accurate with each iteration. And that's all there is to it. It turns out that even a simple one-layer perceptron like the one shown here can learn some simple things . . . and when perceptrons are stacked up on each other, by hundreds or thousands, they can recognize faces, drive cars, and otherwise amaze us.

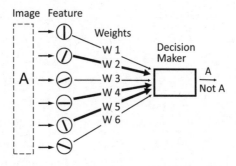

Let's now consider a concrete example. The next picture shows the same perceptron as the first, except with two more feature detectors (the "demons" of the classic pandemonium). The input (the letter A) is presented to all of the feature detectors at once.

Each of these can detect a single feature, in this example an edge slanted at a particular orientation. When we show the perceptron the letter A, the detectors respond as shown: one sensitive to a right-to-left slanted edge responds, as does one that is sensitive to left-to-right, both representing the diagonals on the letter A. In addition, one sensitive to horizontal lines—the crossbar of the A—responds. The teacher tells the decision-maker, "That was A." Accordingly, the A-detecting perceptron increases its confidence in those three detectors.

Notice that, in the diagram, the three lines associated with the letter A have been drawn in thicker lines—weights W2, W4, and W5—to indicate they were weighted as more important. The next time this perceptron is shown a letter, it will pay particular attention to those two diagonal lines and the horizontal one.

Now suppose that we show the A-detecting perceptron the letter B. The demon sensitive to vertical lines would respond, and the demon sensitive to horizontal lines would respond. But the only detector of those two whose opinion meant anything to the A-recognizing perceptron was the demon for horizontal lines. The decision-maker gets much less input than for the letter A, so it responds simply by concluding "not A."

A powerful example of the steps toward perception—by machines and by the brain—is to imagine what happens if we change the size of this perceptron's input. We again show our perceptron a letter A, but this time a smaller one. You will remember that a template-based machine will fail this test, because the small A does not match the original template. But our perceptron will not be fooled, as long as we grant it only one assumption: that our detector can tell the slant of a line in the input image, no matter how big it is or where it occurs in the field. In that case a diagonal line is still a diagonal line, so the feature detectors still find the three features—two diagonals and one horizontal—that define the letter A.

I give this example because the crucial assumption just stated is what the "complex" cells of the visual cortex do. Like my hypothetical perceptron, complex cells of the visual cortex are also sensitive to orientation regardless of where a line occurs within their area of the visual field. This exercise shows how a complex cell of cortical area V1—a fairly simple preprocessing feature detector, present at a relatively early stage in visual processing—can start to build the capability of perception. Yann LeCun, one of the giants of artificial intelligence, explicitly mentions complex cells as an inspiration in his thinking.

BIGGER AND BETTER NERVE NETS

Perceptrons were good fun, but something strange then happened: this type of artificial intelligence (AI) fell into a black hole, from which it only emerged a quarter century later. That gap, from roughly 1965 to 1985, is now termed the "AI winter." During the AI winter, the idea of a learning machine like the perceptron was essentially discarded. We now know that this was a mistake. Nowadays, AI founded on similar principles performs almost human feats. Why did machine learning get shelved?

First, AI was basically an empirical proposal, with not much of a theoretical foundation (that is, a foundation that could be stated in pure mathematics). That was bad. Computer scientists in those days often grew up as mathematicians. Something hard to state mathematically could not be good, they thought. In fact, a leading computer scientist at the time wrote a mathematical proof that a simple nerve net of this kind could not learn anything important.

Clearly, he was wrong, but we didn't learn that from mathematical theory. Rather, it was learned by brute empiricism—by making a computerized nerve net that demonstrably did work.

The second reason machine learning was disdained is practical: computers of the time were by today's standards incredibly slow, and for most scientists, time on them was hard to get. This meant that mathematical theory had to have a big role, because there were few other tools. As computer science developed and as it gained access to fast computers, an element of empiricism, of simple hacking, has crept in. Proof lies not only in mathematical theories now but in results. If nerve nets work, they work; theory, though still important, will have to follow.

Today's large, fast computers and huge training databases have enormously expanded on the basic principle of the perceptron. Here is a canonical diagram of a modern nerve net. The input layer of the diagram simply feeds into a stack of seven perceptrons. These seven perceptrons are wired to a second series of perceptrons, and so forth. In fact, what we understand today as a nerve net is just a stack of concatenated perceptrons.

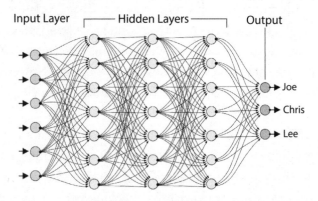

That these are called nerve nets is a bow to the brain. Scientists sometimes refer to the connections as "random," but we know they aren't really, either in the actual machine or in the actual brain; we just don't know the rules of how they work. We do know that the computerized nerve net gets its intelligence from the strengthening of its connections (synapses), which become strengthened during learning. We do know as secure fact that real nerve connections can get strengthened incrementally by repetition of inputs that occur together. And we know that the brain contains lots of layers of neurons.

In machine learning, the great power comes from backpropagation. We saw a simple example in the perceptron. In a modern nerve net, input from the teacher ("this is an **A**") must propagate backward, through each of the hidden layers, modifying connections as it goes, all the way from the output layer to the input—and there can be tens or hundreds of hidden layers. (They are called hidden because they do not communicate directly with the outside world, unlike the input and output layers.) A present task for neurobiologists is to find out how the brain does this—assuming that it does.

———

I HOPE TO be forgiven here by my friends in the machine learning business, for there is much, much more to modern AI than I have shown here. To take a really simple example, you can't just go along forever strengthening a computer's synapses: sooner or later the system would hit a boundary above which the synapses could not be made any stronger. (A solution is to allow negative tweaking of the connections—inhibitory, "anti-Hebbian" synapses.)

But nerve nets can accept any input that can be expressed in terms that a computer understands. They can accept pictures (two-dimensional matrices of pixels), solid objects (three-dimensional matrices of pixels, called voxels), or strings of pressure waves (sounds) if they are properly digitized. They can accept semiliterate chatter on social media pages. A surprising early example was the ability of a nerve net to detect nascent flu epidemics. The input was the ungrammatical chatter on people from various towns' Facebook pages. The "teacher" in this instance was the Centers for Disease Control's (CDC's) report of the instances of epidemics. It turned out that the nerve net, once trained, noticed that a flu epidemic was occurring before it popped up in the CDC's health statistics for a given town. I'm guessing this happened because, among the millions of characters in Facebook chats, there was an increased incidence of health-related posts, including ones using specific terms like "sick" and "day off" and "yucky." There is nothing magical about these correlations; in principle they could have been discovered by a human who monitored the content of these zillion conversations. But we have computers for that, thank heaven; there are far, far too many conversations for a human to monitor.

Just as in the brain, the performance of nerve nets resists degradation. If your computer's nerve net has a zillion connections, losing a small portion of them is likely to affect performance very little. Each connection contains only a fraction of the total "knowledge" owned by that nerve net. To be sure, there will be a subtle loss of function somewhere, but the machine will certainly continue to work. This is just the observation that the surgeon Penfield made about the human brain, much of which can suffer substantial damage with only the subtlest reduction of function.

Finally, and counterintuitively, you don't have to know what's in every hidden layer of a multilayered nerve net. To my knowledge, nobody knows exactly how Apple's nerve net, named Siri, makes the connection between my spoken word and the characters that appear on my iPhone's screen. In principle, these things are knowable: the knowledge is embodied in the set of weights in the synapses of the nerve net. But there are a zillion connections in the net, and tracking down even a single one—the computer's idea of the phonemes (units of sound) that make up the spoken word "dog"—is hardly worth the effort. If the machine works, it works.

TERRY SEJNOWSKI AND A NERVE NET THAT TALKED

During the AI winter, there was a small group of stubborn scientists, led by Geoffrey Hinton in Toronto, who kept on plugging away at the idea of nerve nets. One of them was Terrence Sejnowski (pronounced "Say-now-ski"), who at the time was at Johns Hopkins University. Terry claims that, fortunately, he had not read the influential book proving that nerve nets could not learn anything important. He just went ahead and built one.[1]

He is a brilliant and unusual character. He began his professional life in Princeton's famous Department of Physics. There he published a series of theoretical mathematical papers on the behavior of neurons in the brain. He then went on to a postdoctoral fellowship with Stephen Kuffler at Harvard.

This was a shift for Terry. Kuffler was a hard-core experimentalist, an expert in the nuts and bolts of the nervous system, and he didn't theorize much. What he did have was the acumen to pick Sejnowski from a horde of postdoctoral fellowship applicants.

Perhaps this was because Kuffler recognized in Terry the same sort of essential simplicity that Kuffler himself possessed. If you did not know him very well, you might simply classify Sejnowski as an uber-geek. He loves his science, and seems to work at it night and day. He is interested in everything. And he thinks about things big and small, always searching for a new slant, for the hidden flaw, for something innovative. His thinking is contrarian, in the best sense of the word.

He is also contrarian as a person, of which the most striking example is his preference for white shirts, dark suits, and black shoes. Most scientists think they are individualistic—they wear Birkenstocks, jeans, sweaters, and T-shirts; they drive small cars; and a lot of the men have beards. The clan has its own folkways. In the environment of the neurobiology department at Harvard, a postdoc who came to work in a dark blue suit was clearly someone who followed his own star. I took him sailing once, and he showed up in wool slacks and black leather-soled shoes.

Terry appears strikingly uninhibited in social interactions: he seems to say whatever pops into his head, and this occasionally leads to faux pas, which embarrass him not at all. Plus he has a great, cackling laugh, audible at a distance. The in-crowd of the Harvard neurobiology department didn't know what to make of Terry. He and Kuffler published an interesting and now-forgotten paper on synaptic transmission in a simple model nervous system. But the culture of the neurobiology department, which was heavily empiricist, was not a natural home for Terry. He was viewed as an interesting specimen.

From Harvard, Terry moved on to a faculty position at Johns Hopkins University, and that was about the time he met Geoffrey Hinton, who with David Rumelhart and others had invented the method of backpropagation, which we just saw to be critical for a nerve net's learning—propagating adjustments of its synaptic weights "backward" through each of its hidden layers. Terry grabbed the tool and ran with it.

From my vantage point as a cellular neurobiologist, Terry was pretty silent in the early 1980s. My eyes were opened on a visit to his lab around 1985. There he showed me a neural network that had taught itself to talk.

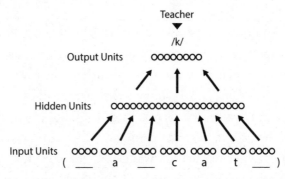

Letters were fed to the neural network one at a time. In this example, the researchers were asking the net how to pronounce the letter *c*, as it occurs in "cat."

Why is this task anything to write home about? Because English pronunciation is notoriously irregular, as anyone who learns English as a foreign language will painfully attest. For example, we supposedly have a rule that vowels are long if there is an *e* at the end of the word, as in "gave" and "brave." Yet the *a* that looks so similar sounds quite different when it occurs in the midst of "have." This violates the rule. And why don't we say the *o* in "mow" like we do in the word "cow"? Native speakers don't notice it because we grow up with it, but computers do. And linguists struggle with the irregularities. They have compiled long lists of rules, exceptions, and exceptions to the exceptions.

Fortunately, they have also created a dictionary of pronunciation. It contains twenty thousand English words, together with their standard pronunciation. Terry Sejnowski and Charles Rosenberg used this dictionary as their nerve net's "teacher." The teacher drew a pronunciation from a dictionary of all the phonemes in English. If the net guessed correctly the pronunciation of the letter *c* in "cat," the dictionary told the net it was right. In the net, the connections between that particular phonetic output and the *c* in "cat" were strengthened.

To make the exercise accessible to bystanders, they (cleverly) fed the computer's phonetic written output into a "speaker," a computer program that takes phonetic written English as its input and turns it into spoken sound. This last step did not add anything to the science, but it allowed a compelling demonstration of the nerve net's learning, one that anyone could relate to.

Before training, the output of the nerve net was, as you would expect, not even a word salad: it was a mush of unrelated phonemes. After a few rounds of teaching, though, it began to speak in something very much resembling baby talk: "ga, ba, ta." A few more rounds of teaching, and the net began to emit some real words, mixed with variously mispronounced ones. And finally it spoke English text almost perfectly—not just the text that it had been trained on, but any English text. And it did this without having been taught any rules of English pronunciation. It had just been shown a lot of examples.

Interestingly, deconstructing a few of the hidden layers showed that the net knew certain word groupings, even though nowhere in its architecture could be found the hundreds of rules of English speech. It is as if the net had learned to speak English in something like the way that a native English speaker does. While only specialists can state the rules of English pronunciation (I certainly can't), native speakers can read English aloud easily, consistently. In learning to speak English, Sejnowski's nerve net behaved the way our human brains do.

All this was done using the computers of the early 1980s—sad and slow things by modern standards. Today, computers are tens of thousands of times faster, and the nerve nets that can now be constructed have hundreds or even thousands of layers. There have been many refinements of the networks, but the root principle is the same one used by Rosenberg, Sejnowski, Hinton, and Hebb.

The talking nerve net was as compelling to everyone else as it was to me. Sejnowski rocketed into the spotlight, where he has remained ever since. He was interviewed on national television. Backpropagation became the default method of refining the connections of nerve nets. Terry soon moved from Johns Hopkins to the beautiful Salk Institute on Southern California's coast, where he has remained ever since.

Sejnowski still wears dark suits. His car is a large black German limousine. At age seventy-one, he retains his distinct, cackling laugh, and despite his many honors and the formalities that go with them, he still conveys a touch of the youthful naïf, of the uber-geek. He is not afraid to talk about the successes of his work, and there are those who begrudge him that, but he is essentially a modest, unselfconscious person, interested in science for its own sake. Despite the great potential for professional envy, I don't know anyone who dislikes Terry Sejnowski.

COMPUTERS THAT SEE

You have probably heard about computers that see. They can steer self-driving cars. They can recognize faces. Science fiction alarmists predict that a camera will photograph your face as you enter Macy's department store, after which a computer will figure out your identity and quickly look up your buying preferences. Macy's will then (somehow) steer you to buy something you didn't come for.

I'm here to tell you not to worry . . . not yet. Remember that CAPTCHAs are still in use as a security device.[2] In fact, the tasks for which CAPTCHAs are used are a handy guide to the things still hard for everyday computers to solve. (The NSA's computers can surely solve most CAPTCHAs, but everyday amateur robots trolling for entry to everyday sites can't.)

The capabilities of computers that see are remarkable, and getting more so with lightning speed. To illustrate, I'll walk us through a couple of ways they do face recognition, the problem I started this book with, the Mount Everest of visual neuroscientists.

It turns out that the best face-recognizing computers are now very good indeed. They are almost as good as humans, though still much bigger and less energy-efficient than a human brain. I'll illustrate two contrasting methods. One is rule-based, meaning that a series of specified analytical steps is rigorously followed. It is the kind of method that comes first to most people's minds—for example, the opinionated aerospace engineer I mentioned at the opening of this chapter. For short, we'll call the rule-based method the "dumb method," although variants of it are far from dumb.

The second method uses machine learning, starts to resemble what brains do, and for now seems to be the method of the future. It is the one that terrifies privacy advocates. We'll call the AI versions the "smart" methods, and I'll concentrate on them, mainly because they resemble what neurons do and I think neurons are smart. At the moment, AI versions dominate the field of face recognition.

The task of a face recognition algorithm has several steps: first to recognize that a face exists, and then to say who that face belongs to. The first task is simply face *detection*, not face recognition. The smart routines and the dumb routines both work with the same starting problem. Given a visual scene containing all kinds of stuff—the men's clothing department of Macy's, for example—they need to see if there are any faces present and isolate them for further analysis.

But even before looking for faces, our computers need to make the test image as clear (to themselves) as possible. These steps, which occur before the algorithm starts trying to identify any faces, are collectively called preprocessing.[3] There are many, many ways to clarify images, as anyone familiar with Adobe Photoshop will attest. I'll take two examples. First, most natural scenes are not uniformly lit: the outdoors sun creates shadows, Macy's spotlights the sports jacket it is pushing that day. For reasons we have already touched upon, we don't notice these variations in brightness, but a digital camera, such as our phone's or computer's, does. This is a problem, because for computers, which are fiercely literal-minded, the same thing illuminated two different ways looks like two different things. Therefore, the first preprocessing transformation of the image is to "flatten" its brightness. The computer averages the brightness of the whole image (sometimes it uses fancier average-like measures) and adjusts the brightness to be consistent throughout the image, as though the image were illuminated by a perfectly uniform light source. Second, most algorithms employ some sort of edge enhancement. We have also touched on this repeatedly; edges are where the action is, and the algorithm sharpens edges to a greater or lesser degree.

Now that our computer has cleaned up the image, the second task is to find faces in it. Once again there are several ways to do this. A method that is fun because it somewhat matches certain neural hardware of the visual cortex is to make a HOG image.

HOG stands for "histogram of gradients." A gradient is just a region that goes from bright to dark—an edge that is bright on one side and dim on the other. In other words, a gradient shows not only there is an edge but also whether the edge is facing inward or outward. The computer measures all the possible gradients in an image, together with their direction, and maps them all.

Such a reduced image is shown here. To make this image, the algorithm's creator broke an image into small squares of 16 by 16 pixels each. In each square, he counted how many gradients point in each major direction (upward, downward, diagonal, etc.). Then he replaced that square in the image with a condensed representation—a simple line—of the edge tuning that was the strongest at that point.

In fact, this image is an average of many HOG images, so it is a composite from many faces, a sort of universal face. You can use this template to locate faces in a crowd. You do this by creating a HOG image for every appropriately sized region of the image—every box of pixels of about the right size to be a face. You move the test box around the scene that you suspect might contain a face.  Then you compare the local HOG images with an idealized example of a face HOG image. Most areas will just yield nonsense scatterings of edges. But a few will match the composite HOG image, and those will be identified as faces by the algorithm. Of course, this does not tell you *who* the face is, only that it is a face. But it allows you to get the face into standard coordinates, an image that you can feed into later steps of your classifier and which is far easier for those later steps to deal with than the raw images. Your algorithm has taken a jumble of pixels—the whole enormous cloud of pixels in a crowd scene—and pulled out of it a selection of pixel arrays that are probably faces.

So we have now cleaned up the image and located the faces in it. A few other adjustments get the faces into a standard format (it is closely cropped, for example, to exclude all the surrounding stuff—that's the thin box you see around faces in some sci-fi movies), and we are ready to try to identify who the face belongs to.

If I go to Times Square and ask random passersby how a face-recognizing computer does it, the majority will say, perhaps in more words: "It has rules that help it tell the difference between facial features. It measures the distance between eyes, for example, or the height of the forehead. It compares those measurements for the unknown face to measurements of known faces."

It is possible to make a face-recognizing computer that works by that kind of fixed rule; an example is an algorithm developed by Matthew Turk and Alex Pentland, which computes something called the eigenvector of the face measurements. At this point in time, however, the vast majority of face-recognizing computers use machine learning instead. This may not be true forever—someone reading this book in ten years may be glad for my warning that rule-based algorithms could make a comeback. For now, though, we'll concentrate on nerve-net-based computers.

Just for fun, let's look at a face recognition algorithm that uses machine learning, in this case a simple one provided by the computer toolbox MATLAB. This will be worth your while because it turns out that a lot of perception works the same way. The first steps, those that help get the face into standard coordinates, are the same as for rule-based routines, described above. Then:

· Feed your cleaned-up (frontal, uniformly illuminated) faces—lots and lots of them—into a multilayered nerve net. The faces on which the nerve net is trained come with labels: "Dick," "Jane," "Bill," etc. This is the machine's teaching step. For each image in the gallery, you tell the nerve net, "This is Bill," or "This is not Bill."

- The nerve net, just like our little perceptron, uses backpropagation to adjust the weights of the individual connections. The synapses that are lit up the most when the teacher says "This is Bill" get strengthened. The difference is that a powerful system uses a whole stack of little perceptrons—these are the hidden layers of AI machines. The backpropagated feedback affects all of the layers, right back to the input.

- Now that our nerve net is trained, we can test it: find an image of Bill and feed it into our nerve net. If the image is enough like one of those in the training set, one of the decision-makers will get a strong input, because its synaptic weights have been strengthened by aspects of Bill's face.

Because the nerve net is very large and has seen a lot of faces, it has become very smart: it can recognize Bill from many angles, brightly lit or dim, in a white dress shirt or a red T-shirt. In real life, the databases used to train face-recognizing nerve nets are huge. In the old days they sometimes used archives of driver's license photos, which could include millions of identified images.

It is fascinating to note that we don't know exactly how the nerve net distinguishes Bill. Skin tone? Ratio of face height to face width? Straight nose or crooked? Dimples? Old acne scars? All of the above? These things lie deep in the hidden layers, with their many thousands of interconnections.

In Chapter 11 we'll explore how much a biological visual system resembles what computers do. I'll argue that malleable synapses are important for visual systems made of neurons at every stage, from the retina to the high places where face recognition takes place. But I'll tell you right now, and in detail in Chapter 13 (spoiler alert), that

MATLAB's nerve net is *not* how object recognition works. In brief, it is too stupid compared to real brains. Perceptrons must have a teacher saying "This is Bill" and "This is not Bill"; this is called supervised learning. Brains can learn without an external teacher—unsupervised learning. We'll come back to this in a big way. But a stack of perceptrons embodies a principle—networks of neurons connected by Hebbian, modifiable synapses—that is likely to be critical in either case.

11 | A Vision of Vision

> Most of the fundamental ideas of science
> are essentially simple. They may, as a rule,
> be expressed in a language comprehensible
> to everyone.
> —ALBERT EINSTEIN

T IS TIME FOR ME TO PUT THINGS TOGETHER, TO GIVE MY ANSWER TO the question that started off this book: How does a parent recognize her child on the playground? To address this is to address a great challenge of neuroscience: the mechanism of object recognition by real brains. The picture I'll give is not the one shown in textbooks, which explicitly or implicitly posits a fixed hierarchy of incremental, additive steps, each step feeding the next until the system reaches a conclusion. Instead, recent work indicates that vision involves, from almost the beginning to the very end, the fluid mechanisms of neural plasticity, learning by nerve net rules.

I should start by creating for you a circuit diagram. Fortunately, I do not have to undertake that massive task myself; Daniel Felleman and David Van Essen of Washington University in St. Louis

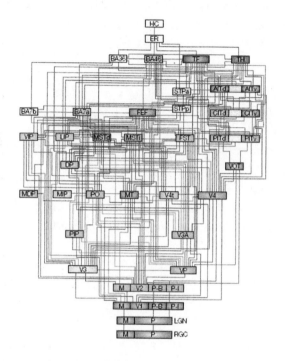

made a map of the connections of the primate visual system. They carefully note that this is a map only of the *major* connections. The rectangular boxes are brain regions. The lines represent axonal pathways among them. We neurobiologists love to show this picture as an example of how terrifyingly complicated the brain is. You might say it's our excuse for having not figured it out yet. Remember, these show only how the gross brain areas are connected. To show all the neuron-to-neuron connections would require millions more lines. At this scale, the whole image would be jet-black.

Where to begin? Let's start with a broad, basic idea about vision, some general principles of how your visual system works. To be sure, experimenters will have to get down into individual connections eventually. Lots of details are not yet known, and higher vision is described in the broadest of brushstrokes. But it will help to start with a concept, a global idea of how the system is designed. Here it is:

I.   Visual systems are not neutral, bias-free recorders of their inputs. At every level, their responses are distorted to match the regularities of the natural environment.

II. In some cases this match is embedded in the genetic code, but in many it is accomplished by nerve net learning. This goes all the way from basic regularities such as sensitivity to edges and lines to complex perceptions such as faces.

III. The coarse connections between the brain's visual areas are made using molecular cues—the kind of cues nature uses during infant development to build a hand or a liver. These are mostly chemical pathfinding signals; they lead axons to target areas in the brain, and help them form a rough topographic map of the visual field in each. But the wiring that underlies perceptions of specific objects—object recognition—is created by neural plasticity.

THE VISUAL BRAIN AS A NERVE NET

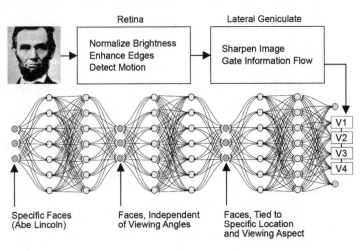

I have already given you the basic experimental facts, inasmuch as they are known. For review:

1. The retina preprocesses the image, taking it apart into a large number of independent representations.

2. The retina projects to the LGN, which sharpens the receptive fields and gates the flow of information that goes on to the cortex.

3. The primary visual cortex, V1, transforms the receptive fields; most V1 neurons respond best to oriented edges.

4. In V1 and V2, many cells gain some positional independence from the original stimulus: there are "complex" cells, which also respond to oriented edges but allow those edges to lie in a wider part of visual space. This represents a step of abstraction, a freeing from the exact visual input.

5. The next cortical areas, V3 and V4, contain neurons with a diverse variety of receptive field preferences—for example, for colors, movements, or visual depth. They project to visual areas located in the temporal lobe.

6. The cortex of the inferior temporal lobe is built of a mosaic of patches, sensitive to different things. Some of the patches specialize in the identification of faces.

7. From posterior to anterior in the temporal lobe the spatial requirement of face recognition decreases, so the more anterior cells are increasingly space invariant—that is, they respond to a face independent of its precise location or orientation in space.

8. To look ahead, in even higher areas, the medial temporal lobe and higher cortices, are cells that respond only when shown the image of a particular person or thing—independent of its location in the visual field or its viewing angle.

Scientists once believed that most of these steps were hard-wired—that the brain was a structure built with fixed connections. But as you have seen, there is now evidence of far greater plasticity. In the sections that follow, I'll walk us through the visual system again, from retina to higher cortex, this time concentrating on the netlike character of its wiring and the plasticity and malleability in learning it displays. You'll see that our natural human vision bears some resemblance to a leading form of computer vision.

THE RETINA

Computer vision usually includes a step called preprocessing or image normalization, which transforms the messy natural image into something simpler and more tractable. This is also one of the things your retina first does with an image: it detects light, and turns the output of its initial light-sensitive cells—the rod and cone photoreceptors—into something the rest of the visual system can begin to work with. First the retina has to normalize output across the huge variations of light intensity present on the earth. This is a much bigger deal than we usually realize. The size of the output of the rods and cones would vary by a factor of one hundred billion from dark midnight to blazing day if their output wasn't normalized by the retina before the brain processed it. Neither individual neurons nor brains nor even computers can deal with that large a range of numerical inputs.

The retina compresses that range so that at any given ambient illumination, the intensity range of its output is only a factor of ten or so. Cleverly, it centers that narrow range on the average brightness of the environment at that time.[1] The only time we are conscious of this process is when we suddenly go from a very dark room into a bright place, or vice versa; we are dazzled or left in blackness until the retina gets itself organized for the new brightness. The second thing the retina does is begin edge detection (by applying contrast enhancement) and motion detection, as we saw in Chapter 4.

What is the point of these early image-processing steps? In computers, a few defined steps begin almost any machine vision algorithm, and the point is to relieve the computational pressure on later steps in processing, be they rule-based analysis or nerve nets. Nature has learned over the ages that things that move are important; the retina embodies that knowledge in its motion-sensitive retinal ganglion cells.

THE LATERAL GENICULATE NUCLEUS (LGN)

In late prenatal life, the axons of the retinal ganglion cells have already reached their target neurons in the LGN. But they do this in an imprecise way: each ending of a retinal ganglion cell divides into many branchlets, and these spread widely to target individual neurons of the LGN. If this continued, our sight would be smeared among the overlapping pathways. Instead, by synaptic strengthening, the eye-specific targeting of the retinal axons is refined.

The way this works in brief is as follows: hardwired molecular cues get the retinal axons to the neighborhood of the LGN, and there they form a crude topographical map. Axons that simultaneously fire their postsynaptic targets—that is, the axons that come from the same eye—have their input to the LGN neurons strengthened. Gradually, wide-spreading axons refine their LGN targets, so one clump of LGN neurons becomes responsive to inputs from the right eye, and another responds to inputs from the left eye. The experiments of Stryker and Shatz that proved this were a landmark, because they involved events that could be verified using exact, reproducible experiments.

THE PRIMARY VISUAL CORTEX (V1)

From here on, you can think of the stages of the visual system as layers in a nerve net. Let me review the stages of object recognition and point out how each could be created by the brain's use of what we now understand as machine learning rules.

The axons of LGN cells project to the primary visual cortex. At that point, neurons sensitive to oriented edges appear. Here's how the cortex builds a simple oriented receptive field from non-oriented LGN fields.

Imagine that we are plotting the responses of a neuron in V1 to a very small spot of light. We get a map of the receptive field that looks as shown on the left below. But these responses to individual small spots are weak; what the cell really likes is an edge oriented along the row of pluses, the excitatory regions. Several LGN cells converge, in a highly specific way, upon the cortical neuron.[2] Groups of these axons from LGN neurons line up such that the LGN neurons' receptive fields on the retina are arranged in rows.

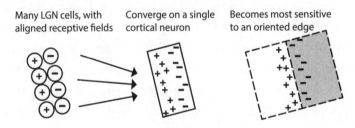

Many LGN cells, with aligned receptive fields

Converge on a single cortical neuron

Becomes most sensitive to an oriented edge

The receptive fields of the LGN cells are the round circles on the left. The LGN axons converge upon a single neuron of the visual cortex. Some of them are excitatory (ON cells, shown as plus signs) and some are inhibitory (OFF cells, shown as minus signs). If the retina is stimulated by an edge oriented at just the right angle (drawing at the far right), all of the excitatory inputs are activated, and none of the inhibitory ones are.

The cortical neuron has an elongated receptive field, with, as you've seen, an excitatory region flanked by an inhibitory region. The optimal stimulus for this cell is as shown in the drawing at the far right: a dark region abutted by a bright region. This is precisely what we mean by an oriented edge. On the light side of the edge, the inputs from four LGN neurons add up, but only when the edge is at the correct orientation.

Lines and edges are important because they are a dominant information-carrying part of the natural scene. This is because of the inescapable fact that our world is made up of objects, and their edges delineate their boundaries; they separate the object from everything else. Often the edges are straight—for example, the trunks of trees. Sometimes they are slightly curved—for example, the edge of a stone. But a curve is just an assembly of little straight lines. So edges are a big part of the input that the visual system receives as it is becoming wired.

It is easy to see how a nerve net, trained by exposure to the natural world, could transform an input composed of round, non-oriented receptive fields into line-sensitive elements—that is, how it could create cortical "simple" cells. Imagine a group of LGN cells converging on a neuron of V1. When a row of LGN cells is stimulated by an edge, those cells fire together and cause firing in the cortical neuron on which they converge. Neurons that fire together wire together, so the synapses of those four LGN cells upon the cortical cell are strengthened, and other synapses upon that cortical cell are relatively weakened.

In fact, this was tested long ago by training a simple computer nerve net, which was simply shown a lot of images of the natural world. (This was a case of *unsupervised* learning.) Its output layer showed that the nerve net had learned to recognize straight lines. That is, at the end of training, the computer contained its version of simple cells.

Remember that a truly random image looks like television snow. Almost any system of vision—natural or artificial—will include edge detection at an early stage.

CORTICAL AREA V2: A COMPLEX CELL

As you remember, a complex cell is orientation selective, like a simple cell, but has a bigger receptive field and cares less about the edge's precise location. It performs a feat of generalization: one could say that it detects "edgeness" for a particular orientation, but is less restricted as to which particular pixels on the retina are stimulated.

The complex cell is thought to be created in just the same way as the simple cell: by convergence of earlier neurons, in this case the simple cells converging on the complex cell. Each simple cell is sensitive to an edge at a particular location. If many simple cells with slightly different receptive field locations converge, the result is a complex cell that responds to edges, but edges spread over a wider space.

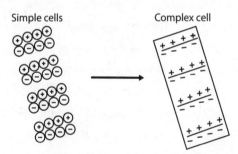

Simple cells   Complex cell

Cortical area V2 contains a lot of these cells. They are also present in V1. I have separated them here to highlight the fact that complex cells seem to be created by a convergence of simple cells in a hierarchical fashion, as proposed by Hubel and Wiesel. This creation of complex cells from simple cells in the brain explicitly inspired a leading form of modern machine learning.

CORTICAL AREAS V3 AND V4

Because of the wealth of connections back and forth among V1, V2, V3, and V4, these areas seem not to be rigidly anatomically sequenced, nor is it easy to imagine a mechanism by which the later receptive fields are built from preceding ones. They are, rather, netlike.

Recording from neurons in areas V3 and V4 reveals cells responsive to a variety of features of the visual input. One that has been studied fairly well is curvature. This was originally described as "end-stopping" by Hubel and Wiesel. These cells are somewhat like complex cells—indeed, it was thought that they were built hierarchically from complex cells—but have the added feature that they like not only edges but edges of a certain fixed length. Subsequent researchers have pointed out that this also can be seen as making the cells curvature-sensitive. But there are lots of other selectivities in areas V3 and V4. Some cells in V4 can discriminate colors. Some cells in V2 are even sensitive to corners. And so the sages cannot yet give me a succinct description of what these cells do.

Following the analogy to computer nerve nets, it may be that V1, V2, V3, and V4 are intermediate-level "hidden layers" of a nerve net. That might be the reason that the behavior of their neurons is—despite major effort—so hard to classify. As you saw in Chapter 10, hidden layers put several nerve nets together in series, vastly increasing their power. They are "hidden" because they do not talk directly to the outside world, only to the next layer. It is not always easy to interpret what hidden layers do, even in a computerized intelligence we ourselves have built. And there is no reason that every neuron in a hidden layer has to do the same thing. We do think the cells in V3 and V4 are more complex in their feature detection than those in V1 and V2. One way to think about them is that these are hidden layers reflecting features intermediate in complexity between those detected by V1 and V2 and those detected by the object-recognizing cells of the temporal lobe.

THE TEMPORAL LOBE

As a loose rule, image processing in the temporal lobe seems to proceed in a posterior-to-anterior hierarchy, with simpler features detected closer to the back, near V1, and complex objects toward the front of the lobe (anterior in the brain). This is a simplification (remember the massive anatomical "circuit diagram" above, with its many feedback loops) but close enough to be useful for thinking about image processing.

There are at least six face areas of the temporal lobe, linked together by axonal connections. They have technical names based on anatomical subdivisions of the temporal lobe, but the labels are hard for non-anatomists to remember (or specialists to agree upon), so I'll take the liberty of collapsing them into a single nomenclature. I'll just talk generally about the posterior, central, and anterior parts of the inferior temporal lobes.

I'll follow current thinking and treat the six face patches spread across these lobes as the hidden layers of a nerve net designed to identify particular visual objects. The core idea here, evidenced by the experiments of Livingstone and her colleagues, whose monkeys' face patches recognized hands instead of faces, is that these layers of the nerve net do not have genetically predetermined jobs; instead, they get trained by seeing faces. They are to some extent, general-purpose recognizers.

If face-recognizing computers are a guide, the brain's hidden layers in the posterior and central temporal lobe detect increasingly complex collections of facial features. The hidden layers of the posterior temporal lobe take input from V1–V4 and use it to recognize things like faces, noses, chins, hairlines, and especially eyes. Intuitively it is possible to imagine how the complex selectivities of preceding layers—selectivity for curves, corners, and so on—might allow parts of faces to be defined. This sounds like magic, but it is actually less magical than it seems, because the pixels of a face do not occur at random. But how this is done is not known in mechanistic detail. As is often the case in nerve nets, the exact computations of the hidden layers are obscure.

The posterior and central temporal lobe face patches seem to assemble facial features into a simple representation of a face—a "proto-face," if you like. Such a face uses a constellation of features it inherits from earlier hidden layers. There is evidence from Tsao and her colleagues that these features are relatively simple ones, such as the aspect ratio of the face, distance between the eyes, and so forth. The pixels of a face do not occur at random. The two dark holes representing nostrils occur more often in pairs than by chance, and they also occur more frequently above the line of pixels that define a mouth than by chance. The individual elements that define a face become linked—if you will, they make a cell assembly.

These cells are sensitive to images of real, biological faces, but they are easily tricked by symbolic faces—ovals containing dots for eyes, short straight lines for noses and mouths. In fact, Tsao has shown how the individual elements can be combined mathematically to yield a representation of an image's degree of "faceness." And she has played around with this, observing that if you show one of these neurons a simple face that is missing one eye, it responds quantitatively less strongly than to an image with both eyes intact. Yet even here, the face must fall in a specific place in the cell's receptive field, just as simple cells of V1 require an edge in a particular place.

It appears that the central and posterior parts of the temporal lobe feed forward to this next layer, the anterior temporal lobe, which is a space-invariant face recognizer. Here, many cells recognize a face, with some latitude as to where the face falls visually and its exact array of pixels. The mechanics of this have not been specified in detail, but they are very likely to resemble those of large face-recognizing computer networks, which share this ability. Some cells can recognize the image of a face, and also the mirror image of a face, no matter where in the visual field the face lies. Other cells achieve true space invariance, responding to a face no matter where in the visual field it lies. Why mirror images are important is not clear. A likely interpretation is that the mirror-image-recognizing cells are a step—a hidden layer—toward true space invariance.

Finally, and even more remarkable, there are cells specific for the faces of particular individuals. The brain appears to have cells and circuits—parts of cell assemblies—that allow us to recognize all the people we know from daily life: our friends, family, colleagues. Apparently, the space-invariant nerve net's output is input to a representation—even higher in the chain of perception—that learns to recognize specific individuals. But how such cells would integrate into the whole system is a matter of pure speculation.[3]

In summary, then, imagine the stages of temporal lobe vision as a series of five kinds of events. First, the brain's nerve net learns to recognize face parts. Second, the brain's idea of a face is assembled from its computation of face parts—noses, eyes, and so on. Third, neurons fire in response to the image of a face if it is located at a particular location within its field. Fourth, some cells achieve partial view invariance. Finally, neurons in the most anterior face patch achieve almost full view invariance. And in humans a nearby area—one of the targets of the anterior temporal lobe—contains cells that respond to only a small subset of face identities. Thus it appears that a major goal of the face patches is to build, in stepwise fashion, a representation of individual identity, of people or of things.

You will perhaps notice a certain vagueness about the details of this postulated nerve net. That is because we are far from a mechanistic understanding of higher visual processing, one that is based on specific neurons and their synapses. Indeed, there are ways in which it is already clear that the brain cannot use the simple perceptron-like nerve nets used by computers to recognize faces and drive cars. To anticipate the discussion later, most of these use supervised learning, while the brain must use some form of unsupervised learning. I wish here to emphasize the general principle more than any particular form of a nerve net, which is that object recognition works by multineuronal assemblies, formed by the progressive modifications of synaptic strength, as Hebb suggested.

It is important to remind ourselves that faces are not the only thing recognized in the temporal lobe. There are other patches concerned with many other types of objects, linked visually or conceptually. A nice example is the cells that respond to images of tools—not any particular tool, but tools as a category (hammers, saws, pliers). We are only beginning to unravel the logic of the temporal lobe.

PART III
TO THE HORIZON

As you probably have noticed, this book has progressed from pretty solid facts to less interpretable ones. Now we are going to lean over the edge of the cliff, barely touching secure earth. Why get into that dangerous posture, otherwise called speculation—a known peril to be shunned by serious neurobiologists? The reason is that otherwise I have told you a shaggy dog story, a tale with no ending. There is no way to avoid bumping into the questions asked in Part III. What comes next in the chain of perception? Nobody knows for sure, but I'll try to show you the terrain, that gray land where perception merges into thinking.

12 | Why Evolution Loved Nerve Nets

> It is not the strongest of the species that survive,
> nor the most intelligent, but the one most
> responsive to change.
> —CHARLES DARWIN

WE HAVE TO ASSUME THAT, GIVEN A FEW MILLION YEARS, NA-ture can build anything it wants. Why nerve nets? The short answer is that it was more efficient for evolution to design a modifiable synapse than to redesign the genome—the genetic plan for the visual system—for every animal species. Machine learning—by brains—is what you might call a broad-spectrum mechanism. It can adapt to the characteristics of different visual environments (near vision in the forest, far vision in the plains), and the same mechanism can allow you to recognize your child on the playground.

Its first advantage concerns the building of the brain during early life. It may help here to imagine the alternative: a brain made up of fixed connections, a recognition machine that looks like a beautiful eighteenth-century Swiss chronometer, all gleaming brass. In this brain, each person to be recognized would have to correspond to a particular combination of gears and wheels. Obviously, there is no way that such a machine could encompass all the individuals to be recognized. And remember that all of those people would have to be programmed into the machine in advance, before the machine had seen any of them. To return to the human machine, our building plan—the genetic code that makes a fertilized ovum turn into a person—would have to be specified for each possible individual to be remembered. This is ridiculous: nature does not know in advance what individuals the brain will have to recognize.

It is also ridiculous in terms of the amount of coding that would need to occur. The primary visual cortex contains very roughly fourteen million neurons that project to V2, which in turn contains roughly ten million neurons. If the system were precisely determined, each axon tagged for a specific V2 neuron, axon guidance would require far more molecular signals, each coded by a particular DNA sequence, than there are genes in the genome (a paltry twenty thousand). Nobody supposes that this level of specificity really exists. Even with clever shortcuts, the genome is orders of magnitude short of containing the amount of information necessary for perception.

So instead of trying to specify all of the connectivity, nature uses a dual strategy. Programmed rules for neuron guidance (molecular highways, diffusible signals) lead axons to the correct spatial region: face patches are in about the same place in all primates. They also cause the topography of each region to retain the spatial map of the retina. But this is true only at a crude level. The precise connections that code for specific objects are determined by machine learning rules after the crude layout has been formed. In Livingstone's phrase, molecular mechanisms create "proto–face patches." They gain their gross location by rigid, genetically programmed developmental cues, and their final selectivity by synaptic plasticity.

Second advantage: nerve nets overcome the problem of having to recognize all of the remembered objects in a person's visual history, seen from all possible angles of view and distances. In a nerve net, a given neuron can participate in several recognizers, depending on how its output is further processed by the next layer. If there are many layers in the nerve net, and if each layer contains, let's say, a few tens of thousands of elements (neurons), the number of possible combinations is astronomically large. This creates a huge combinatorial advantage—enough to encompass the face of your grandmother from many views as well as your child and all the other children on the playground, running and jumping at a thousand different angles.

Putting all this in the most general terms, the overarching advantage of nerve nets is that *they allow the brain's visual system to be matched to the natural scene* in which the animal lives. The nerve net that serves visual recognition is trained on precisely the visual objects around us. The simple case is oriented edges, which are a primary feature of almost all visual scenes in a mammal's life. But certain complex objects are also important. For example, faces are critical to social animals such as primates, including us. If a test monkey did not get to see faces, the "face areas" became dedicated to something else—hands in the example we saw. The systems for recognizing edges or faces or hands are *learned* by their brains.

———

I CANNOT RESIST here pointing out the elegance—the economy and simplification—of nature's design for the brain's sensory systems. We have seen over and over a great organizing principle of sensory systems: sensory systems are tuned to the statistical regularities of the natural world, the features of the visual input that are most important for the animal. For a few components of the visual world that are always present—for us and for our evolutionary ancestors—it made sense for evolution to spend the genes that cause the retina to be tuned to those aspects. Again, the detection of contrast (edges) serves as an example; this is a prewired function, created by molecular instructions evolved over eons, from horseshoe crabs to humans.[1]

Lateral inhibition is simple enough to be biologically preprogrammed before birth, but for complex objects there are too many tunings to accomplish with our limited set of genes. Nerve nets have no such constraint. By Hebb's rule, the brain can become matched to any higher-order regularities that exist. Eyes, noses, chins, and hairlines do not occur at random in a young monkey's visual world. They occur together, the prerequisite for perceptual learning.

Here's another example. Consider the shape of the quork in the following illustration. Now imagine a world in which the "faces" of the important players—Martians, perhaps—looked like quorks. In that case, the Martians' temporal lobes would have a series of cortical patches that detect quorks, because quorks were the visual image upon which their nerve nets had been trained.

On Earth, however, we humans would still have temporal lobes with human face patches. Even though one of us might occasionally see a quork, it would not make sense for many of our neurons to be devoted to quorks, as compared with faces. You may never see my drawing of a quork again, but you surely will encounter thousands of faces, and they will be critically important for your life. Your visual system is not a general-purpose recognizer. You spend neurons on things that you actually see, not on things that you might see.

To be sure, this way of thinking about vision deals with the very practical matter of efficiency in neural hardware. But there is also a certain elegance, because it means that the brain of an individual animal is linked in the very wiring of its cells to the natural world in which that animal lives. Because of perceptual learning, our visual brains contain a built-in copy, a distillation, of our natural environment.

13 | Some Mysteries, Some Progress

> If you don't know where you are going,
> you might wind up someplace else.
> —YOGI BERRA

THE FACE CELLS IN PRINCIPLE SOLVE THE CLASSIC PROBLEM OF OBJECT recognition, the one given to you in Chapter 1: some cells in the anterior temporal lobe respond to a face regardless of the angle, the lighting, and the position on the retina. But how far have we gotten, really, toward understanding what most people think of as perception?

Not so far, really. For brains, we run out of steam at the level of a few neurons in the temporal lobe, neurons that can tell you a particular object is in the neighborhood. A pretty skimpy perceptual world, that. And our other favorite models, seeing computers, are good only at very narrow tasks, and are clunky and inefficient. Let's look again at those computers, where despite flamboyant successes a huge, challenging problem has not yet been solved.

SUPERVISED AND UNSUPERVISED LEARNING

We hear daily about the perceptual tasks that can be accomplished by machine learning: self-driving cars, face recognition, and much more. There is also wringing of hands over the dangers of intelligent machines let loose, the risk that they will somehow take over the world, using their intelligence to leap ahead of humans. In most of these discussions, however, there is an elephant in the room: machine learning can indeed do some remarkable things, but, as the AI folks are painfully aware, it is still way dumber than the average four-year-old.

The reason is that the most famous AI algorithms were taught to do their trick only by using vast amounts of data and superfast computers. A four-year-old learns things by him- or herself, often after just a few examples. To be sure, we painstakingly teach certain skills and concepts to our children. But most of what they know— their basic mechanisms of perception—are self-taught; their little brains do it all by themselves. In fact, even a multilayer perceptron, backpropagation and all, is far dumber than my little grandson, who did not need a zillion examples and a teacher to learn to recognize his grandfather. Just a few hugs and he quickly learned to say "Granddad."

AI researchers distinguish between supervised learning and unsupervised learning. You'll remember that our simple canonical perceptron has a teacher. So does Apple's voice recognition software, Sejnowski's talking computer, and the face recognition software that frightens privacy advocates. The face recognition software is taught using huge catalogs of faces together with their identifying labels. Computers can do this because they are so fast—indeed, the triumphs of recent machine learning have come about substantially because of the very recent (past five years) availability of huge training data sets and large, purpose-built computers. Because their neurons work so slowly, brains can never compete in those departments.

But somehow they do compete, and more. Hebb's original concept recognized that cell assemblies were created by *unsupervised* learning: the contiguous points of an edge tend to occur together all by themselves, and the brain creates edge-sensitive neurons all by itself. A pressing current task for computer scientists is to make machines more brain-like in their training.[1]

Soon I want to get back to wet brains, but first, an example of another form of machine learning, one that combines nerve nets with a new principle. Computer scientists have generously named these strategies after something from neuroscience: the principle of reinforcement, first systematically studied by the great Russian physiologist Pavlov and elaborated by descendants like B. F. Skinner of Harvard. They call these algorithms "reinforcement learning." Reinforcement just means that a certain behavior gets rewarded, and if it is rewarded, it tends to get repeated. The steps that led to the final correct behavior become more likely to be repeated; the synapses in its nerve net get strengthened. If you will, this is a form of backpropagation, and reinforcement learning is like perceptron learning except that the computer generates its own teacher.

Computers can be rewarded, too. In reinforcement learning, the computer is told to seek a goal. It throws out a guess, initially a pretty bad one. But if its guess will take it a tad closer to its goal, the computer is rewarded. The computer does not get a pellet of Purina Computer Chow. Instead it gets told, "Good job. Increase the synaptic weights of the things you just did." Then it goes around again for another try, this time starting with the new weights. You can imagine the rest: it goes round and round, tweaking its weights each time, until it learns to do its task.

But reinforcement learning has already mastered one astoundingly hard task: playing chess, or even playing the harder game Go.

Not only are these computers playing at superhuman levels—they can beat any human—but they *taught themselves how to play*. The one I am thinking of is called AlphaZero, reported in a paper in *Nature* just before Christmas of 2018. It was given only the rules of the game: a definition of the board, the rules by which each piece can move, and so on. It was then set to playing chess or Go against itself. Sounds a bit nonintuitive, but the trick is that each self is not allowed to know what the other self is thinking—only to know what move the other self made on the board. There is no teacher, only some internal rules that tell the computer how to know if a move was a good one or a bad one, and finally whether it won the game. In four hours the computer was playing world-class chess.

This is an amazing feat, applicable not only to board games but to many other tasks as well. In one of his lectures, David Silver of Google's AI team shows a video of AlphaZero using a remote control to make a toy helicopter fly aerobatic stunts. Watching the helicopter do barrel rolls is enough to make a believer of anybody.

But is AlphaZero as smart as my grandson? Not by a long shot (unless my grandson challenges it at chess). The computer needs a very narrowly defined task. And its brain is a whole lot bigger than my grandson's and needs more than Cheerios for power. Katherine Wu estimates in *Smithsonian* that AlphaZero's hardware runs at a power of about 1 million watts; my grandson's brain uses less than 20 watts. The point of bringing up both nerve nets and reinforcement learning (remember that AlphaZero incorporates a nerve net in its internal machinery) is just that they are proofs of principle— proofs that a certain kind of logic can generate performance that approaches, even distantly, that of a brain.

Could the human brain implement the manipulations involved in a deep nerve net or AlphaZero? Of course, though prohibitively slowly. The human brain is a computer designed by millions of years of evolution, with triumphs of miniaturization in its synapses and their connections. If these things can be done by a bunch of computer chips—cumbersome objects by comparison—they can be done by brains.

The people in the AI community are well aware that my grandson beats their computers, and they are making progress. How smart the computers will get is anybody's guess. I think they'll get pretty damned smart—I would be the last person to bet against them. There are fascinating variants of unsupervised learning in the works. The only question is how long it will take, and how much the machine solution will resemble the solution used by real brains. And, importantly, will the computer solution ever approach the brain's in economy of hardware? I'm not holding my breath for that one. In fact, the amount of energy it would have to be fed is what keeps me from worrying, for now, about a superhuman computer that would take over the world.

WINFRIED DENK AND THE CONNECTOME

For sheer naked creativity in neuroscience it is hard to beat Winfried Denk, director of the Max Planck Institute of Neurobiology in Martinsried. If you are aiming to emulate him, though, you have to know that his particular form of creativity only occurs between midnight and 4:00 a.m.

Winfried is a tall, solidly built man with shaggy hair and a small beard. He is usually seen smiling. Google shows only one picture of him without his smile, and in that picture he is wearing a white shirt and a necktie (rare for him, as he usually goes about in jeans and loose shirts—maybe that's why he isn't smiling). The occasion was the awarding of the Kavli Prize, one of the big honors in neuroscience; apparently, this required a dignified formal portrait.

That prize and others were awarded for a remarkable series of important discoveries, most of them based in Winfried's training in physical science and optics. The first was the development of the confocal microscope. The confocal microscope, which was mentioned earlier, is an optical instrument that offers greater resolution than would have been possible before. Confocal microscopy melts together optical microscopy and computer analysis—there is no image in the traditional sense, just a series of scanned spots that are later reassembled into an image digitally. It rapidly took over and is now the industry standard for microscopy.

This was work he did at Cornell as a postdoc, with his advisor Watt Webb. Confocal microscopy had been around, at least in theory, for a while: what Winfried did is to make it practical for biologists. But the next innovation, two-photon microscopy, was thought up and patented by Winfried and David Tank during a stint at Bell Laboratories, the famous creativity factory once supported by the former telephone company and now, sad to say, victim to the vagaries of the business world. Two-photon microscopy lets you see more sharply, deeper into the tissue, and with less damage to the tissue than ever before.

Winfried then moved back to Germany, to head the Max Planck Institute for Medical Research in Heidelberg. With more resources, he could hire his own stable of machinists, engineers, and programmers and could spread out a little, working on not one but two big projects.

The first, done in collaboration with junior faculty member Thomas Euler, was to use the two-photon microscope to solve the half-century-old problem of direction selectivity in retinal neurons. Euler, Denk, and their colleagues used the two-photon microscope to image one of the retinal amacrine cells—the starburst cell—while stimulating the retina with moving things. (They could do this only because of the two-photon microscope, which illuminates the tissue with a wavelength of light that barely stimulates the rods and cones. Otherwise, the retina would have been unable to respond to the test stimulus, because it would be blasted by the usual microscope's illuminating beam.) The cells of these retinas contained an activity indicator, so the investigators could see the starburst cell responding to light. Lo and behold, the starburst amacrine cell, which synapses upon the direction-selective ganglion cell, is itself direction selective: without going into the mechanism here, it causes the direction-selective ganglion cell to be direction selective, too.

While solving direction selectivity, however, Winfried was behind the scenes developing yet a new innovation, and this one didn't even use optics. Quite the opposite, in fact—it was a way to run around the problems of optics. That one may have bigger consequences for neuroscience.

Before delving into that area, which has been termed connectomics, it is worth taking a look at Winfried's modus operandi. I have already said that he works at night—a good way to buy undisturbed time. But what is he doing when he's in his office at 2:00 a.m.? One thing he is not doing is the mundane chores of a professor—preparing lectures is not required by the Max Planck Society, and he avoids reviewing submitted manuscripts and other chores. As to running his laboratory, he is a skilled delegator. He ignores some of the time-sucking stuff and is a genius at delegating the rest. He reads, and he thinks—something most of us do only in spurts—carefully and deeply.

Another thing that he does, surprisingly, is to spend a lot of time on airplanes. He gets invited to lecture a lot. However, he does not particularly enjoy the speaking part of the job. The invitations are mainly a chance to schmooze. Winfried is a three-star schmoozer, and his schmoozing has a method. He gravitates to folks who specialize in science near one of his interests, and then he relentlessly probes their minds. I have more than once heard him say, when learning of a possible new source, "Hey, I had better visit him soon"—even if the person is a complete stranger, on the West Coast, or in China.

I helped him get interested in the problem of direction selectivity, and for a few years I could expect a visit once every few months.

After his big paper solving the problem was published in *Nature*, he came to visit me again, and he remarked: "You gave me my last problem. What do you have for me next?" But I didn't have a new problem, at least none that would reach the importance of direction selectivity. His visits tailed off after that. Please understand that this is not a criticism of Winfried. His visits benefit both the visitor and the visited—they are the best of scientific communication. I'll schmooze with Winfried anytime.

He could give hundreds of talks a year if he accepted all the invitations. Winfried's talks are unusual, because they are—or at least seem—almost unprepared. He wanders up to the stage, looks at the floor, and seems to mumble whatever comes into his head first. The talks seem unrehearsed, casual, and sometimes wandering. He forgets which slide comes next. He is far from inarticulate in person and his written work is crystal clear. But his talks are a far cry from the slick presentations favored in the present era, where people spend days practicing their lines and polishing their graphics. Winfried always has something important to say, but he does not lead you to it by the nose. I honor him for flouting that time-wasting part of the current enterprise.

Now for the connectome. The *-ome* ending of the word gives it the meaning "all the connections," and the connectome is a breathtaking attempt to learn just that: all of the connections between the brain's neurons. For sheer chutzpah, it is hard to beat. It may not be accomplished in my lifetime, or Winfried's. But Winfried has shown how it can be done, and the juggernaut that is scientific progress will take care of the rest.

How do you identify a connection between two neurons? You do that by identifying a synapse between them, and that's not easy, because synapses are small—roughly 0.5 to 1.0 micrometer. At that size, you have to use electron microscopy, and there's the rub. As you have already heard, traditional electron microscopy requires sectioning the neural tissue into thousands of ultrathin slices, each around 50 millionths of a micrometer (50 mm) thick. That means you need hundreds or thousands of serial sections to span even a single synapse, much less the distance between two cells. Using traditional methods, no human could cut enough sections to span that distance, and even if you could, how would you get the sections properly aligned with each other?

Winfried invented a solution to those two problems. It is called block-face scanning. It is actually a form of electron microscopy, but with a couple of twists. The first is that you don't save the section once it has been cut; you throw it away. What does that leave? Your tissue block, with the cut face, the surface from which you just cut a section, staring you in the face. Winfried's innovation was to look at the cut face instead of the section, collecting images using something called a scanning electron microscope. Because the tissue block remains stationary, the whole cutting process can be automated. And the problem of alignment of the images is mostly solved, because the cut face of the block is in pretty much the same place each time a section is cut.

Thus Winfried ends up with a long series of images, representing serial sections through the tissue sample, at the resolution of electron microscopy. After that, there are many technical obstacles to solve, but suffice it to say these images can be aligned and neurons can be traced through the series, across even moderate distances (for now, and we hope longer distances in the future). In each section there may be hundreds or thousands of neurons, so the tracing is a big chore (a hard one even for computers). But in principle, the final result includes *all* of the connections among the neurons.

Why is that such a big deal for neuroscientists? Because that's all there is. The brain is a connection machine; if you know all of the connections, you are well on the way to understanding how the brain works. Winfried tried this out on the retina and successfully confirmed the synaptic connections that make direction selectivity happen, giving the answer a solidity never before accomplished for any neural circuit.

Since its invention, the technique has been modified and improved, and a number of labs are using it for different problems and with different variants on the method. But the big advance was the conceptual one: to aspire to an analysis of all of the connections that make the brain do its thing. There are still technical things to work out, and it will be a while before long-distance connections—even those between neighboring brain regions (i.e., V1 to V2)—are amenable to this kind of attack. Even then, there will remain important questions to answer: What kind of neurotransmitters are used at the synapses? What messages do they carry? But it surely will happen. This much is certain: in the end—the very long run—the connectome will be the foundation of any understanding of complex neural circuits.

What will the man do next? What new and heretofore unsuspected innovation will come to Winfried Denk in the early morning darkness? We don't know, but the record shows that it will be something wonderful.

NERVE NETS IN SIGHT

Many of the star players in machine learning come from computer science—they are three-quarters computer geek and one-quarter neuroscience geek—but the explorers of intelligence still include plenty of people who like to get their hands dirty, people who want to know, for sure, how the brain works, on the tangible evidence that you only get by messing around with actual brains.

The first question that comes to a career experimentalist like me is "How the hell can we study nerve nets, made of tens of thousands of neurons spread all over the brain? Record from thousands of neurons at once? Even if you could do it, how would you digest the data?" A decade ago it seemed impossible, but things are getting better.

As usual, we needed to put together advances in disparate areas. Four of them were particularly important. The first was the two-photon confocal microscope, which lets you see way more sharply than a traditional one. And it lets you see not just the surface of your object but down into it; in the case of the brain, you can see right through the layers of the cortex. As you just learned, the two-photon version was pioneered by Winfried Denk, one of the truly creative minds in all science and (I hope) a future Nobelist. Someone has written that the difference between traditional and two-photon microscopy is like watching a color television in darkness versus with bright lights on. And two-photon microscopy allows us to observe in a way that does not damage the cells (an issue with traditional confocal microscopy).

The second advance was genetic engineering. This allowed us to engineer a protein into the brain's neurons—a protein that flashes (more accurately, changes fluorescence) when a particular neuron is active. If you look at a piece of nervous tissue engineered with such a protein using a confocal microscope, you can actually see, in real time, the activity of individual neurons.

The third advance was an import from the world of insect biology. Suppose you want to track the way a beetle runs—you want to see, for example, which cues make it turn left and which make it turn right. You could just take a movie of the beetle running and then hire a graduate student to score the beetle's movements. Instead, insect biologists figured out an automated way. You glue the beetle's shell to a stationary platform so that the beetle's legs are now dangling in the air. But you position underneath the beetle a lightweight sphere, like a Ping-Pong ball, which the beetle now grasps firmly with its legs. The Ping-Pong ball is suspended in an almost frictionless carrier, so when the beetle runs, the ball now spins. You measure the turning of the ball, which you can do with a computer. (Your graduate student can now go on to something more interesting.)

The fourth advance is one we moderns take for granted: cheap computing power. When the two-photon microscope shows a few thousand cells, each flashing its little activity-driven light, you have on your hands Big Data. Without modern computing power, we experimentalists couldn't make sense of this.

The final thing needed was a neurobiologist who is brilliant, persistent, and brave. He is David Tank, from Princeton University, co-inventor of two-photon microscopy, who put all these pieces together and then added some cool wrinkles of his own.

"Let's aim for the stars," thought Tank. "Let's try to see thousands of neurons at once, in a conscious and undisturbed animal, and see them while the animal is looking at things and thinking thoughts about them." Tank (with others) figured out ways to hold a mouse still in a restraining frame, like the beetle in our example. The mouse doesn't mind; it gets fed there. The mouse stands on a freely moving ball, again like the beetle. Tank points a confocal microscope through a skull window at the mouse's cortex. This is a mouse that has had an activity indicator introduced into its cortical neurons, using genetic engineering methods. So the mouse can do its natural behavior—running—while Tank and his colleagues monitor the activity of the brain neurons. And, oh yes, the mouse looks at a synthetic world, created using the technology of virtual reality. Tank and his colleagues can then teach the mouse various tasks—running down a virtual maze, for example—while watching how its neurons behave.

This technique is pretty new, but it works. And the fundamental finding is that neurons behave from day to day in consistent ways: for example, in most cases the same group of primary sensory neurons seems to light up each time a particular stimulus is shown. It didn't have to be that way—they could just blink away according to their own logic, in patterns incomprehensible to us. Indeed, neural responses in cortical areas tasked with sensory-motor integration behave in more fluid ways, sometimes drifting, sometimes stable, which have not yet been entirely deciphered. This is not unexpected, as these areas represent connections between sensory input and behavioral actions, which vary situationally. But this is a substantive, conceptual challenge, not a technical impediment.

That's as far as I'll go in describing this line of work. Tank's students are spreading the technology throughout neuroscience. Mice are running down virtual mazes, their neurons blinking signals into the experimenters' computers. Your brain and mine boil over with experiments that could be done. Could we watch a cell assembly get made? Do memories stay in the same place in the brain, or do they migrate? What do neurons do when the brain sleeps, when the mouse wants something, when it sees a mate? The tools are there, and all that remains is to do the experiments. Stay tuned: they are happening as we speak.

14 | In the Distance

Somewhere, something incredible is waiting
to be known.
                              —BLAISE PASCAL

I know not all that may be coming, but be it
what it will, I'll go to it laughing.
                              —HERMAN MELVILLE

ONE GREAT QUESTION LOOMS BEHIND EVERYTHING THAT HAS
been said here about perception: Who is the looker, and where
does he or she "stand" in the brain? It is natural for us—scientists
and civilians alike—to think of perception as we ourselves try to understand it. We stand outside the brain (or the computer) looking
inward, seeing nerve impulses or electrons running to and fro, and
trying to deduce the ways in which those things reflect the outside
world. And that is an achievable task: to figure out how neurons of
the brain represent the real physical world. Represent it they do—
they have to in order to avoid predators, even to walk down the sidewalk. Our brains make a faithful map of visual reality, one that allows
us to pick up objects or fly through the trees on our skis. But it is not

fair to postulate a little human inside the head, looking at the sensory input like a viewer at a movie.[1]

Maybe the most obvious difficulty with everything I've said about vision is what's known as the binding problem. How do the variegated visual inputs to the brain get bound together to create a unified perception—the single perception of a red automobile moving toward the right?

The decomposition of visual input into parallel streams of information is a fundamental principle of the brain's visual system. Let's think again for a moment about the retina. The first tasks of the retina are to detect light and to compress its response to a narrow working range of brightness. But much of the retina is devoted to fragmenting the visual image, breaking it into separate signals representing movement, color, edges, and the like. These signals are truly separate: a retinal ganglion cell that detects the direction of motion does not tell the brain much about the color of the moving object, and a cell coding for color does not say much about the presence or absences of edges.

These disparate signals go to different places in the brain. The most numerous cells project straight on to the LGN and thence to the visual cortex, but some of the others split off to serve other functions. Some cells detect absolute brightness and send a signal to brain centers concerned with the sleep-wake cycle. Others provide a specialized signal used to stabilize the eyes. But nowhere do those signals come back together anatomically—if anything, they are further separated into cortical simple and complex cells, complex movement detectors in MT, and so on. How do they get "bound" again into a single perception?

Neuroscientist Anne Treisman's proposal in answer to this question relies on the fact that disparate areas of the early visual system all seem to contain a topographical map of the visual image: V1 has such a map, V2 has a similar map, and so does MT, and so do the higher areas of the temporal lobe. What if the topographic maps were linked so that the cells representing the color of the image, say, mapped point for point onto the simple or complex cells in areas V1, V2, V3, V4, and the face patches of the temporal lobe? Area-to-area connections of that sort are known to exist.

The image on the right might come from a collection of boundary-enhancing neurons in V1—a bit like a HOG image. The middle image might be the output of a neuron that does edge enhancement only, like an LGN neuron or one in the retina. And the third image might combine those two with a representation that conveys color, or shades of gray (like the sustained cells of the retina). If all three of these images were present in separate places in the brain as separate topographical maps, the brain might then reunite the pieces by superimposing the three maps, resulting in the picture on the left.

A different type of answer, championed by Wolf Singer, Christoph von der Malsburg, and others in Germany, is that the coherent image of the object is tied together by synchronized firing of neurons in the separate representations. Neurons representing an object would fire in response to the edge-enhancing image. They would also fire in the color-analyzing image. If they fired in synchrony, that would indicate to the rest of the brain that these two neurons belong to the same object. Debate continues about the plausibility of this mechanism.

A RELATED AND harder problem is ancient: Where or what is our consciousness? Where or what is the "self"? Most of us feel as though our existence is centered in the head, somewhere not too far behind the eyes. Could it be that "self" that solves the brain's binding problem? That self could look at all of the fragmented signals arriving from the retina and somehow paste them back together to yield an image of a single object. But how is that person who inhabits my head, the center of my personal universe, generated by a lump of soft brain tissue? It's not just me: all of us have the subjective feeling that we exist. But what or where is our being? I am not satisfied by any of the answers proposed thus far.

For example, the Turing test is an exercise famously proposed as a test for machine consciousness. You make the best computer that you can, and you train it to simulate the thinking of a person. Then you stand outside that computer and engage it in conversation. If you cannot distinguish the computer's conversation from that of a flesh-and-blood human, Turing proposed, then that computer is conscious. I'm afraid that I've always thought that the Turing test was, like the emperor in the fable, not fully dressed. Turing was clearly a brilliant mathematician, but the Turing test has nothing to do with mathematics. Exactly *why* does the perfection of this ideal computer's simulation mean that it is conscious? It is still nothing but a hunk of silicon.

Hebb thought that our personal identity—our self—was the concatenation of all our cell assemblies, not only the simple ones used for perception but also the grander ones incorporating thought, memory, emotion, and the myriad of other things that make us human, all interlinked in a single brain. That concept at least has the virtue of concreteness, and I have some sympathy toward a notion that incorporates all our past experiences. Given what we now know about specialization in the brain, Hebb's model would need a bit of modernization—his big cell assembly would have to possess little specialized regions, corresponding to the task-specific patches that exist all around the cortex. Hebb did have a concept, though, a vision of how a whole brain might work as a series of nerve nets.

We have now traced visual perception to its station in the temporal lobe, where there are neurons that respond to particular faces (or to other classes of objects). We assume that these face cells are embedded in cell assemblies, because one neuron by itself is unlikely to accomplish much. What's next?

More cell assemblies, of course. Hebb did not think there were bright lines between perception, thought, and action—the last being the neural commands that direct our muscles to move. All of them were represented by overlapping nerve nets. The figure illustrates his concept. (Hebb will probably spin in his grave—so let's call this "neo-Hebbian.")

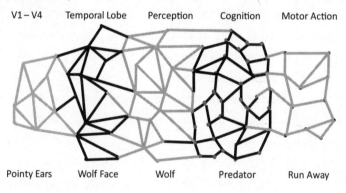

The dots in this diagram represent neurons, the lines between them axonal connections. Note that the individual functional regions overlap: the cell assemblies for perception interpenetrate with those for cognition. Moving from left to right we see, first, the early stages of cortical vision. These areas, V1–V4, do a preliminary analysis of the visual image. We know a fair amount about V1, where cells become sensitive to oriented edges, although there are hints of more complex selectivities. It has been pointed out that V1 does things that help image "segmentation," the parsing of the visual panoply into discrete objects. Cells in V2 elaborate, responding to oriented edges but with less dependence on exact localization. V3 and V4 cells are hard to give a pithy description of. They have varied selectivities, defying even the most skilled experimenters to give a unified account of the region. As a summary of the group of these areas, it seems pretty clear that cells in V1–V4 are selective for particular features of visual input, but lots of them: edges, corners, curvature, color, and almost certainly other things that single out elements, other features, of the image.

Paths through the network represent cell assemblies, formed during perceptual learning by Hebb's rule. They can be short, enclosing few neurons in small polygons, or large, crossing the boundaries of the different functional areas. Any path through the whole network is allowed. The V1–V4 cells make connections back and forth with the visual areas of the temporal lobe. Note that there is no sharp boundary between the cell assemblies of the primary sensory areas and the temporal lobes. A cell assembly located mainly in the temporal lobe, for example, might overlap one or more cell assemblies of the primary visual areas. In fact, that is how early features become connected with higher-order representations. When the right set of primary features is activated, those learned connections activate the more general perception—the wolf's face in this example.

These merge with cell assemblies that represent abstract thoughts, only loosely connected with perception. In each case the nerve nets overlap. Because of the overlap, the different aspects trigger each other. Most of the neurons of the "perception" net are connected primarily with neurons in that same net, but some are shared with the "thoughts" nerve net. Because of that overlap, some of the perceptions can initiate thoughts—and, in fact, be part of them (hence we can imagine sensory events). In other words, a thought can trigger activation of a "perception" nerve net so that we recall perceptions. Some of the "thoughts" cell assemblies overlap with the "actions" nets, so that they can trigger activation of those cell assemblies and, finally, bodily movements.

Thus, cell assemblies for simple visual attributes merge into higher-order representations, always directed by connections learned from simultaneous activation originating in the real world, by Hebb's rule. Cell assemblies for concepts merge into assemblies for actual motor actions. This is what I meant by saying that there is no bright line in Hebb's thinking between perception and action.

This is of course a cartoon, designed only to make the essence of Hebb's concept concrete. And as previously noted, it would have to incorporate recent attempts to assign specific functions to very local regions of the brain. But note that the multicell imaging tools described in Chapter 13 allow deeper and deeper exploration of neurons in their native networks, and they show events more consistent with the kind of picture shown here. That picture is also more faithful to the brain's actual anatomy, that daunting array of feedforward and feedback connections, both between cortical areas and with subcortical centers.

Now let's look at the same sequence in the language of a modern nerve net—a more or less random network of connections, with an input layer, several hidden layers, and an output layer. This is again a cartoon, an abstraction, aimed only at expressing a few basic conceptual features of the algorithm.

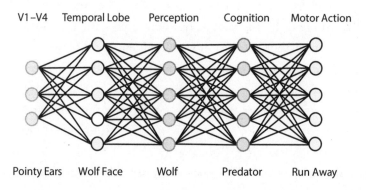

V1–V4 Temporal Lobe Perception Cognition Motor Action

Pointy Ears Wolf Face Wolf Predator Run Away

You will recognize this as the same sort of arrangement we cooked up to recognize Abe Lincoln on page 173, except that we have extended it beyond the face-recognizing part to include perceptions, thoughts, and actions. This nerve net is a general-purpose one. It recognizes the wolf, but also many other things, all depending on which synapses got strengthened during its learning phase. Just as in Hebb's proposal, there are no bright lines between the layers, because the chain of synapses can be modified by experience from one end to the other—by backpropagation or whatever mechanism turns out to be used. We know that the brain's nerve net acts as its own trainer: it is an unsupervised learner. Perhaps it uses reinforcement learning, or one of the other unsupervised methods now under discussion.

———

THAT'S AS FAR as I want to go here: a concept, vague in most particulars, that the steps from visual input to action proceed by a series of modifiable neural connections—cell assemblies or nerve net layers. Aside from the practical advantages—we know from computers that such a network can do some fairly smart things—what does it teach us? The most important thing here is that these models do not have a central decision-maker, a little human receiving signals and sending out orders. It is, like the real brain, a network of connections.

But where does this take us on the search with which this discussion started, the search for the neural identity of the self, the "me" that lives inside my head? Assume for the moment the correctness of the view shown in these two alternative pictures. Can we find consciousness in them? Sad to say, no. In fact, they are ultimately behaviorist models: sensation leads to perception, which leads to thought, which leads to action. They don't have to be physically separate; they can exist within parts, even distributed ones, of a nerve net. You could almost imagine it as a chain of responses, like Ivan Pavlov's conditioned reflexes: the sight of food leads to the idea of food, which leads to the dog salivating. But a chain of reflexes does not have to be conscious. Such a system is like the machine that fills bottles in a Coca-Cola factory, operating by chains of actions: fill the bottle, put on the top, stick on the label—and most people would deny that this machine is conscious.

To further complicate things, it is clear that a whole lot of what our brains compute is not conscious. Where does the solution to a math problem come from? When you ride a motorcycle, are you constantly saying to yourself, "Lean toward the inner radius of the turn"? Perhaps you could verbalize that last rule (and perhaps not), but you don't have to play it over and over as you zoom through the countryside. The famous neurosurgical patient identified as H.M. in the literature lost the ability to form any new overt memories after having brain surgery to control his epilepsy; he could read the same magazine over and over, not knowing that he had read it before. In effect, his conscious life ended the day of his surgery. However, he could still learn simple motor skills, though he was entirely unable to say anything about them and did not himself know that learning had occurred. Once again, this documents the difference between conscious learning and unconscious learning.

Hebb did not get deep into consciousness. He thought that it somehow resided in the widespread activity of the cerebrum, many phase sequences all active together. Christof Koch, who has thought a lot about consciousness, feels that "consciousness is a fundamental, an elementary, property of living matter." And he accepts the necessary corollary, which is that "any and all systems of interacting parts possess some measure of sentience." This leads, among other places, to the question of animal consciousness. Is a dog conscious? Probably. Is the worm *C. elegans*, which has only 305 neurons, conscious? Only a tiny bit. But how do we decide about the nerve net of a jellyfish, or the 135,000 neurons in a fruit fly? Hebb, Koch, and many others recognize that consciousness in mammals happens only when distributed cerebral systems are working in a coordinated way; consciousness goes away when the more primitive centers that control the cortex are damaged or shut the cortex off (as in sleep). So everybody agrees that consciousness *depends* on certain large brain structures. But what neural circuits contain consciousness, and what activity in those neurons makes consciousness happen? Perhaps you could postulate that some superordinate phase sequence, crossing all the boundaries of the brain's nerve net, is where consciousness lives—some sort of emergent property. But that "explanation" says almost nothing at all; it only renames the problem. Perhaps the problem of consciousness is a trick of language, like Zeno's paradoxes. Perhaps it is even a new property of matter, like mass, as Koch seems to suggest. But the intuition of our consciousness has no handles, no analogies, nowhere to stand and look at the problem. It is intrinsically subjective, contained in one person at a time. In the end I fear that consciousness is unknowable. "When I see an apple," G. E. Moore asked fellow philosopher Bertrand Russell, "do I see the same red as you?" To my knowledge, nobody yet has given a convincing answer.

Glossary

acetylcholine: A molecule used widely in the nervous system as a neurotransmitter for communication among neurons, and from neurons to muscle cells.

action potential: A brief, all-or-nothing electrical signal within a neuron. Also known as a *spike*. Action potentials are propagated down axons to signal from neuron to neuron.

alpha cell: A structural and functional type of retinal ganglion cell. Alpha cells have ON transient or OFF transient responses to light.

amacrine cell: A type of retinal interneuron. Amacrine cells receive inputs from bipolar cells and from other amacrine cells, and make outputs onto bipolar cells, other amacrine cells, and ganglion cells.

axon: A thin process that constitutes the output element of a neuron. In most cases, signals run from the cell body to the axon to a synapse upon another neuron.

backpropagation: A method for adjusting the weights of the connections present when a nerve net is instructed by a teacher.

beta cell: A type of retinal ganglion cell. Beta cells have smaller receptive fields than alpha cells. They have ON sustained or OFF sustained responses to light.

bipolar cell: A type of retinal interneuron. Bipolar cells are the through-conducting neuron of the retina, receiving input in the outer retina from photoreceptor cells and making output in the inner retina to amacrine and ganglion cells.

cell assembly: A group of brain neurons that become synaptically connected due to simultaneous activation.

circadian clock: The brain's system for maintaining a circadian rhythm in bodily functions. An example is the twenty-four-hour sleep-wake cycle.

complex cell: A type of neuron in the visual cortex. Complex cells respond best to a line or edge of a specific spatial orientation. The line/edge can be present anywhere within the receptive field of the cell, as long as the orientation is correct.

cortex: The outer covering of the brain's hemispheres. The cortex is connected to subcortical structures by many reciprocal axonal pathways. The cortex is required for sensory perception and higher mental functions.

dendrite: Classically, the input element of a neuron, though it is now known that a dendrite can in some cases be an output element as well. Dendrites are thin extensions from the main cell body of a neuron.

direction-sensitive cell: A neuron that signals to the brain the direction in which a stimulus moves across its receptive field.

dopamine: A chemical used as a neurotransmitter at many synapses in the brain and retina.

edge enhancement: An image-processing routine that enhances the local difference in signal intensity at the boundaries of different objects in the image.

electroencephalogram: A recording of the brain's activity obtained from electrodes placed external to the brain, usually on the scalp.

electron microscope: A microscope in which electrons rather than photons (as in a light microscope) are focused on a sample. Because electrons have a shorter wavelength than photons, the resolution of electron microscopy is higher than light microscopy.

fluorescence microscope: A microscope that images the emitted fluorescence of a sample. The fluorescence is stimulated by irradiating the sample with bright light at one wavelength; it causes emission of light at a different wavelength.

Hebb synapse: A synapse that behaves as postulated by Donald Hebb, which becomes strengthened when the pre- and postsynaptic cells are simultaneously excited.

hippocampus: A brain structure of the limbic system, favored by experimenters because of the stereotypy and accessibility of its neurons.

HOG image: A histogram of gradients image. A map showing the locations of brightness gradients and their directions in an image.

horizontal cell: An interneuron of the outer retina. Horizontal cells receive inputs from rod and cone photoreceptors and make outputs back onto the rods and cones and to bipolar cells.

immunocytochemistry: A method for locating proteins within brain or other tissues. It relies on the availability of antibodies to the protein in question.

lateral geniculate nucleus (LGN): A nucleus located in the dorsolateral thalamus, the main way station for visual information passing from the retina to the cortex.

long-term potentiation (LTP): Persistent strengthening of a synapse based on recent activity.

magnetic resonance imaging (MRI): A technique that produces computerized images of internal body tissues, based on nuclear magnetic resonance of atoms within the body induced by the application of radio waves.

Moore's law: Roughly, that the computing power of the current generation of transistors doubles every eighteen months.

MT: Middle temporal cortex. A region of visual cortex specialized for detecting stimulus motion.

neurome: A listing of all of a neural structure's cell types.

neurotransmitter: A chemical used to transmit excitation or inhibition across synapses.

photofilling: Various techniques by which irradiation of a cell by light is used to cause a diffusible marker to fill the cell, including its axon and dendrites, rendering it visible among the unstained neighboring neurons.

photoreceptor cells: The rod and cone cells of the retina. These are the neurons that are sensitive to light.

postsynaptic neuron: The neuron that receives the output of a synapse.

receptive field: The area of a sensory surface from which a particular neuron can be excited or inhibited. The original definition was purely spatial (a "field"); more recently it includes the other properties of the necessary stimulus, i.e., movement sensitivity, bug detection, etc. One speaks of a "direction-sensitive receptive field."

redundancy: A term taken from information theory to describe overlap of information within an information source. In vision, for example, redundancy means that a point in the natural scene is more likely to contain the same characteristics as neighboring points, rather than different characteristics.

retinal ganglion cell: A type of retinal neuron that receives input from other retinal neurons (bipolar and amacrine cells) and makes output to the brain. The bundled axons of the retinal ganglion cells make up the optic nerve.

rods and cones: The two types of photoreceptor cells in mammalian retinas.

simple cell: A cortical neuron that responds best to a line or edge at a particular orientation. Distinguished from a complex cell by requiring that the edge fall in a narrow region of the retina.

spike: *See* action potential.

sustained cell: In sensory systems, a neuron that continues to respond as long as its adequate stimulus is present.

synapse: The point of signal transfer between two neurons. Synapses can be excitatory or inhibitory. The word is used as a verb as well as a noun, e.g., "The bipolar cell synapses upon the retinal ganglion cell."

transient cell: In sensory systems, a neuron that responds at the initiation of its adequate stimulus, but responds less to the continued presence of the stimulus than to its onset.

Acknowledgments

I am grateful to my editors, initially the estimable Beth Raps and Lisa Ross. I met Lisa, representing SDP Publishing, at the Harvard Publishing Course, to which I am also grateful, and she directed me to Beth. Beth made spectacular improvements. Every time I thought I had things looking pretty good, she made them better. At Basic Books, Eric Henney took a different tack: where Beth concentrated on the words, Eric concentrated on the ideas—the implications of perception. He pushed me to stretch beyond the realm of hard-core biology, and the book is more interesting for it.

I owe much to my colleagues at Harvard and MIT for their seminars and discussions. They know that I have been writing a book, and many have read individual chapters. Out of respect for their openness, I have made it a policy not to write about any of their work that has not yet been published. I am also grateful to the retina community worldwide, my friends and critics. For comments on individual chapters, I thank the following, who have more than once saved me from error: Judith Ames, Mark Ames, Richard Born, Chinfei Chen, Philip Craven, Don Donderi, Marla Feller, David Ginty, Christopher Harvey, Gabriel Kreiman, Margaret Livingstone, Steven Massey, Peter Sterling, Enrica Strettoi, Uygar Sümbül, Roy Wise, Jeremy Wolfe.

Special thanks to Elio Raviola, my friend and collaborator, who still works in the lab at age eighty-five. Thanks to Terrence Sejnowski for an advance copy of his book. Special thanks also to Gerald Shea for encouragement and good advice at an early stage. This book does not follow the usual popular science format and I am grateful to Jim Levine of the Levine Greenberg Rostan Literary Agency for taking a chance on it. He thought that the project might have merit and steered me to Basic Books and the folks mentioned above.

Last but far from least, I thank my wife, Jean, for her thoughtful comments on the manuscripts and for her understanding when I was distracted, but mostly just for being herself.

Figure Credits

Images not specifically credited below are by the author and Haobing Wang.

Chapter 1. *Three faces:* Detail from a New York City Ballet promotional image. Photo credit: Paul Kolnik.

Chapter 4. *Drawing of bipolar cells:* Image by Elio Raviola.

Chapter 5. *Basketball player:* Retrieved from Pixabay@pexel.com. Image by Keith Johnson. *Diverse retinal neurons:* Image by Richard Masland and Rebecca Rockhill.

Chapter 6. *Simple cell:* Adapted from Hubel, D. (1988). *Eye, brain, and vision.* New York: Scientific American Library. *Complex cell:* Adapted from Hubel, D. (1988). *Eye, brain, and vision.* New York: Scientific American Library.

Chapter 7. *Lateral view of the macaque brain:* Adapted from Yang, R. Visual areas of macaque monkey. Retrieved from http://fourier.eng.hmc.edu/e180/lectures/visualcortex/node6.html. *Face areas in the marmoset brain:* Hung, C. C., Yen, C. C., Ciuchta, J. L., Papoti, D., Bock, N. A., Leopold, D. A., & Silva, A. C. (2015). Functional mapping of face-selective regions in the extrastriate visual cortex of the marmoset. *Journal of Neuroscience, 21*(35), 1160–1172. *Deconstructed face:* Tsao, D. (2014). Detail from: The macaque face patch system: a window into object representation. *Cold Spring Harbor Symposia on Quantitative Biology, 79,* 109–114.

Chapter 10. *HOG image:* From Geitgey: https://medium.com/@ageitgey /machine-learning-is-fun-part-4-modern-face-recognition-with -deep-learning-c3cffc121d78. *English-pronouncing nerve net:* Redrawn from Rosenberg, C. R., & Sejnowski, T. (1987). Parallel networks that learn to pronounce English text. *Journal of Complex Systems, 1,* 145–168. *A canonical nerve net:* Adapted from Bengio, Y. (2016). Machines who learn. *Scientific American, 314,* 46–51. Original graphic by Jen Christiansen.

Chapter 11. *Connections in cortex:* Felleman, D., & Van Essen, D. (1991). Distributed hierarchical processing in the primate cerebral cortex. *Cerebral Cortex, 1,* 1–47.

Chapter 14. *The binding problem:* Sinha, P. (2014). Once blind and now they see. *Scientific American, 309,* 49–55. Image courtesy of Project Prakash.

Notes

CHAPTER 1: THE WONDER OF PERCEPTION

1. Vision uses around one million fibers running down the optic nerve. The auditory nerve has only thirty thousand. The two senses represent different things. Vision represents, first and foremost, space—the distribution of inputs across your retina. Hearing represents time, the order in which pressure waves reach your ears. Interestingly, the temporal resolution of the auditory system is way, way greater than that of the visual system. But the problem of segmentation is the same in both cases: for vision, to separate distinct objects from the surrounding clutter, and for hearing, to separate a sound from those that came before or after.

CHAPTER 2: NEURONS THAT SING TO THE BRAIN

1. Some of them excite the postsynaptic neuron, and others inhibit it. Some act very quickly (milliseconds), others slowly (seconds or tens of seconds). In many but not all cases, a single neuron contains only a single chemical species of neurotransmitter. Given that individual types of neuron have thousands of different patterns of connection with other neurons, the added diversity of neurotransmitters adds a huge variety to the computations that the brain can accomplish.

2. In fact, there is a complete map of the body's surface on the surface of the brain; body parts are represented by locations on the brain. When the "thumb" area of the brain is activated, the brain knows that the thumb has been stimulated.

3. For example, a pattern of spike firing analogous to the Morse code for the letter A could tell the brain that the particular fiber is a quickly adapting touch afferent.

4. David Ginty. See Zimmerman, A., Bai, L., & Ginty, D. D. (2014). The gentle touch receptors of mammalian skin. *Science, 346,* 950–954. Abraira, V. E., & Ginty, D. D. (2013). The sensory neurons of touch. *Neuron, 79*(4), 618–639.

5. The profile of Kuffler comes from the author's personal recollections and from essays collected in McMahan, U. J. (1990). *Steve: Remembrances of Stephen W. Kuffler.* Sunderland, MA: Sinauer.

6. This is not terribly easy to do, which accounts for a prominent paper concluding wrongly that hawks have astoundingly high visual acuity. This turned out to be due to a flaw in the behavioral paradigm used to test hawk vision. See Gaffney, M. F., & Hodos, W. (2003). The visual acuity and refractive state of the American kestrel (Falco sparverius). *Vision Research, 43,* 2053–2059.

CHAPTER 3: A MICROPROCESSOR IN THE EYE

1. This mechanism and its early discovery are well described in Dowling, J. E. (2012). *The retina: An approachable part of the brain.* Cambridge, MA: Harvard University Press.

2. Gollisch, T., & Meister, M. (2010). Eye smarter than scientists believed: neural computations in circuits of the retina. *Neuron, 65,* 150–164.

3. Ames Medium, almost a half century after its invention, is item number A1420 in the 2018 catalogue of the Sigma-Aldrich chemical company.

4. Boycott, B., & Wässle, H. (1999). Parallel processing in the mammalian retina: The Proctor Lecture. *Investigative Ophthalmology and Visual Science, 40,* 1313–1327.

CHAPTER 4: GHOST NEURONS

1. Enrica had started out with the thought that a so-called enriched environment would be good for vision, as it is for various kinds of mouse intelligence. Once she found that it was beneficial, she did more experiments in which each component was studied independently. Barone, I., Novelli, E., & Strettoi, E. (2014). Long-term preservation of cone photoreceptors and visual acuity in rd10 mutant mice exposed to continuous environmental enrichment. *Molecular Vision, 20,* 1545–1556.

2. It is only fitting that Elio Raviola would be a master of the Golgi technique, as he received his own training in the Department of Anatomy at the University of Padua, which had been Camillo Golgi's own department in the nineteenth century.

3. In the decade since this work was published serial-section electron microscopy has allowed the subdivision of two of Wässle's bipolar cell types, so the actual number is estimated by some at fourteen types. This is partly a matter of semantics, and Wässle's fundamental point is correct. Wässle, H., Puller, C., Müller, F., & Haverkamp, S. (2009). Cone contacts, mosaics, and territories of bipolar cells in the mouse retina. *Journal of Neuroscience, 29,* 106–117. Helmstaedter, M., Briggman, K. L., Turaga, S. C., Jain, V., Seung, H. S., & Denk, W. (2013). Connectomic reconstruction of the inner plexiform layer in the mouse retina. *Nature, 500,* 168–174.

4. This profile of Boycott comes from the author's personal interactions and two excellent biographical essays: Boycott, B. B. (2001). Brian B. Boycott. In Squire, L. R., (Ed.). *The history of neuroscience in autobiography.* San Francisco: Academic Press. Wässle, H. (2002). Brian Blundell Boycott, 10 December 1924–22 April 2000. *Biographical Memoirs of Fellows of the Royal Society, 48,* 51–68.

CHAPTER 5: WHAT THE EYE TELLS THE BRAIN

1. As a simplification, I have been writing here about thirty representations—thirty types of retinal ganglion cells. In fact, this is just a convenience so that I don't have to keep adding species-specific qualifiers. The actual number appears to vary in different species, which live in environments that put different demands on their vision. In mice, the number might be as high as fifty. In primates, it seems to be lower.

Too, different species have other specializations, notably in the distribution of cells across the retina. Many ground-dwelling animals have a "visual streak"—a horizontally oriented swatch of high neuron density. This helps prey animals scan the horizon for predators. Primates have a tiny region of the central retina in which a special, miniature type of ganglion cell is packed exceedingly densely. That's why human vision is so much better in the center of the visual field than in the periphery, as was noted early on in this book. But all retinas work according to the same fundamental plan. All retinas carry out the same operations of brightness normalization and edge enhancement. And all retinas studied thus far fragment the image into parallel representations, each version telling the brain something different about the image.

2. Stevens, C. F. (1998). Neuronal diversity: Too many cell types for comfort? *Current Biology, 8,* R708–R710.

CHAPTER 6: SENSORY MESSAGES ENTER THE BRAIN

1. These are of course only the main places. Numerically small projections of the retina go to as many as fifty targets in the brain. The best known of these are the pretectal nuclei, which generate tracking eye movements. But there are many more, sometimes of unknown function.

2. The most recent physiological evidence suggests that in the mouse many LGN neurons receive input from a single functional type of retinal ganglion cell, but that others receive input from mixed types. Whether this is specific to the mouse remains unclear. Roman Roson, M., Bauer, Y., Kotkat, A. H., Berens, P., Euler, T., & Busse, L. (2019). Mouse dLGN receives functional input from a diverse population of retinal ganglion cells with limited convergence. *Neuron, 102*(2), 462–476. Rompani, S. B., Mullner, F. E., Wanner, A., Zhang, C., Roth, C. N., Yonehara, K., & Roska, B. (2017). Different modes of visual integration in the lateral geniculate nucleus revealed by single-cell-initiated transsynaptic tracing. *Neuron, 93*(4), 767–777.

CHAPTER 7: WHAT HAPPENS NEXT

1. Don't worry too much about the exact locations of these areas; a general impression will do. Among other things, the surface contours of the brain are remarkably variable—at least as much as, say, the shapes of people's noses. Also, experts disagree about the proper naming system.

2. There are some differences between Tsao's idea of the mechanism and others. Tsao believes, with some direct evidence, that the brain measures a large number (fifty) of parameters about the face—inter-eye distance, for example—and then combines them to get a distinctive signature of that particular face. A more common view at present is that the mechanism is less deterministic, as will be described in Chapters 10 and 11.

CHAPTER 8: THE MALLEABLE SENSES

1. Karl Lashley (1890–1958) was a pioneer of the discipline of neuroscience and thought deeply about the relationship between brain structure and behavior. He carried out a long series of experiments in which he sought to pin down the connection between brain regions and memory, the

so-called memory trace, which Semon had dubbed the "engram." Lashley was Hebb's thesis advisor and longtime friend, and his work was basic to Hebb's theory. Lashley's work is summarized in his classic essay: Lashley, K. S. (1950). In search of the engram. In Society for Experimental Biology (Ed.), *Physiological mechanisms in animal behavior (Society's Symposium IV)* (pp. 454–482). Oxford, UK: Academic Press.

2. During the 1970s, controversy raged over these experiments. The argument was essentially between nativists—Hubel and Wiesel—who thought that the wiring of receptive fields was innately programmed, and environmentalists, who thought, with Donald Hebb, that the wiring of receptive fields was strongly influenced by visual stimulation. Hubel and Wiesel published a paper reporting that orientation selectivity in monkeys was present at birth, contradicting Hebb's view that lines were detected by learned cell assemblies. I once saw a curt—not to say rude—letter from Hubel to Hebb saying so in no uncertain terms. A study in the macaque monkey later reported that some cells of V1 in some species have orientation selectivity in animals too young to have received significant visual input, or in animals raised without the experience of viewing any contours, and this led to a downplaying of Hebb's proposal. However, Hubel, who was right about so many things, appears to have it wrong in this instance. It was soon learned that it is a minority of cells that show such an orientation preference, more in some animal species than others, and even those cells are far less sharply tuned than normal. See Espinosa, J. S., & Stryker, M. P. (2012). Development and plasticity of the primary visual cortex. *Neuron, 75,* 230–249.

CHAPTER 9: INVENTING THE NERVE NET

1. Author's personal recollections and interviews conducted in Chester, Nova Scotia, August 2016. Biographical information on Hebb may be found in Hebb, D. O. (1980). D. O. Hebb, In Lindzey, G. (Ed.) *A history of psychology in autobiography.* San Francisco: Freeman. Brown, R. E., & Milner, P. M. (2003). The legacy of Donald O. Hebb: more than the Hebb synapse. *Nature Reviews Neuroscience, 4,* 1013–1039.

CHAPTER 10: MACHINE LEARNING, BRAINS, AND SEEING COMPUTERS

1. Material comes from the author's personal recollection and from Sejnowski's history of artificial intelligence: Sejnowski, T. (2018). *The deep*

learning revolution: Artificial intelligence meets human intelligence. Cambridge, MA: MIT Press.

2. "CAPTCHA" stands for "completely automated public Turing test to tell computers and humans apart." A CAPTCHA consists of distorted images designed to be easy for humans to understand but hard for computers. Traditionally these were distorted letters, but recent ones often use pictures instead—for example, a collection of pictures of a bus, occluded by various other objects, with the instruction to count the buses. It is a race between the hackers and the CAPTCHA designers.

3. In the computer vision world the features detected here are sometimes called "priors." They are biases from prior experience about important classes of environmental events. You can simply give your machine the priors, or it can discover them using machine learning. Interestingly, it was once thought that it was always better to reduce the amount of computer work needed by giving it the preliminary features ahead of time—we know what many of them are, so why not give them to the algorithm for free? Counterintuitively, however, there is now evidence that it is more efficient for the algorithm to learn the early features on its own, by machine learning.

CHAPTER 11: A VISION OF VISION

1. In outline, wide-field retinal interneurons—horizontal and large amacrine cells—sample the brightness over a wide region of retina. They then "subtract" a factor depending on that brightness from the signal transmitted down the photoreceptor to the ganglion cell path. There are other mechanisms, some working on a long time scale and some rapidly, some working in the outer retina and some in the inner. For example, in bright light the response of the photoreceptor cells is desensitized. In the inner retina, amacrine cells play a role in directly adjusting the ganglion cell response.

2. The basic mechanism described here was postulated by Hubel and Wiesel and has been directly observed in paired recordings from LGN and V1 neurons. Reid, R. C., & Alonso, J. M. (1995). Specificity of monosynaptic connections from thalamus to visual cortex. *Nature, 378,* 281–284.

3. The famous paradigm is the "Jennifer Aniston" cell, encountered in recordings from a human patient during surgery. It fired only when the person was shown a picture of the actress, and not when other movie

stars were shown. Of course, Jennifer's is not the only face that can be recognized by this person's brain—she is just the one whose image the experimenters happened to hit upon. And in truth, we don't know what to make of this finding. Presumably the Jennifer Aniston cell is one cell of a larger neural network, but we have no clue what kind of network it might be.

CHAPTER 12: WHY EVOLUTION LOVED NERVE NETS

1. The horseshoe crab *Limulus* is found, in a form identical to modern horseshoe crabs, in fossils from the Paleozoic Era. Edge enhancement (lateral inhibition) was first discovered in the eye of the horseshoe crab. Even though its "retina" is far simpler than that of mammals, the same type of edge enhancement operates: illumination of one photoreceptor causes inhibition of its neighbors.

CHAPTER 13: SOME MYSTERIES, SOME PROGRESS

1. This is an understatement. Computer scientists forthrightly point out that a young child is a better learner than their computers. Some of those scientists are purely task-focused: they want only to build an AI that will distinguish a good credit risk from a bad one. But others, the dreamers, have bigger game in sight. They want to build a general intelligence, one that will at least compete with my grandson. This may be the "singularity," the master algorithm that could take over the world. I do not dismiss the futurists' scenarios and I'm glad that these thinkers are at work. But I'm not worried myself . . . not yet. The main reason is that I see the master algorithm as limited not by software—at which the AI people are demonstrably very good—but by hardware. As noted in the text, even AlphaZero, with all its awesome capabilities, is still pretty much of a one-trick pony, good at certain tasks, impossible at others—and it is an energy-hungry monster. Imagine a version that could excel at every possible intellectual task! Extrapolating from the current version, this would be a computer as big as the Ritz. Are there that many megawatts in all of North America?

What is beyond doubt is that machine learning will soon do some pretty sophisticated tasks, for better or worse. I'm an optimist and think for better. In the meantime, there is a fine ferment in the AI field, well worth paying attention to.

CHAPTER 14: IN THE DISTANCE

1. In this chapter I give you one example of a problem to which one can imagine at least the class of a solution—the brain's binding the fragmented parts of the visual image. It is only a physical problem. The different representations exist in some physical form or the other in the brain so that you can imagine links between the different features, or some sort of neural cross-signal, that tells the brain, "All this stuff belongs to the same object in the world." This is only a problem of finding the links.

Consciousness, the self, is a problem of a different order. The contents of my consciousness are known only to me; it is a subjective thing—or is it even a "thing" at all? Could it be only a trick of language, like Zeno's paradox? I think not, because everyone is quite certain that they have one. We cannot deny that it exists. But what it actually *is* remains a mystery. Quotes are from Koch, C. (1982). *Consciousness: Confessions of a romantic reductionist*. Cambridge, MA: MIT Press.

Bibliography

GENERAL INTRODUCTORY READING

Ackerman, D. (1995). *A natural history of the senses.* New York: Vintage. "The senses feed shards of information to the brain like microscopic pieces of a jigsaw puzzle. When enough 'pieces' assemble, the brain says 'Cow. I see a cow.'" Ackerman is a poet and brings to the senses a poet's eye. An elegant book, containing accurate science and the poet's-eye-view of it.

Dowling, J. E. (2012). *The retina: An approachable part of the brain.* Cambridge, MA: Harvard University Press. Classic textbook, which has introduced generations of students to the retina. Good introduction to the responses of the retinal interneurons. Revised in 2012 after its initial publication in 1987.

Hubel, D. (1988). *Eye, brain, and vision.* New York: W. H. Freeman. A review of Hubel's work, at the level of *Scientific American.* The writing is elegant and packs a punch; there is thinking behind every word. Don't try to read it too fast.

Masland, R. H. (2001). The fundamental plan of the retina. *Nature Neuroscience, 4,* 877–886. An attempt to cut through species differences to the organization basic to all mammalian retinas.

Rodieck, R. W. (1998). *The first steps in seeing.* Sunderland, MA: Sinauer. Bob Rodieck had an astounding breadth of knowledge about vision, from the details of phototransduction to the anatomy of retinal ganglion cells to the psychophysics of color vision. Even though a bit out of date, this is a treasure trove of facts and thinking about vision. Beautifully illustrated by Rodieck himself, with help from his longtime assistant Toni Haun.

Wolfe, J. M., Kluender, K. R., Levi, D. M., Bartoshuk, L. M., Herz, R. S., Klatzky, R. L., & Merfeld, D. M. (2017). *Sensation and perception*, 5th ed. Sunderland, MA: Sinauer. An engaging and authoritative textbook, covering vision and the other senses.

ADDITIONAL READING

Much of the best material is found on the web, especially in the area of machine learning. For example David Silver, a leader of the team that built AlphaZero, now posts a lucid series of lectures (his course at University College London) on YouTube. But these will likely be gone, or be different, in five years. I cite websites as the URL stood when they were encountered. The reader's best strategy would be to Google "machine learning lectures" and follow where it leads.

Abraira, V. E., & Ginty, D. D. (2013). The sensory neurons of touch. *Neuron, 79*(4), 618–639.

Afraz, A., Boyden, E. S., & DiCarlo, J. J. (2015). Optogenetic and pharmacological suppression of spatial clusters of face neurons reveal their causal role in face gender discrimination. *Proceedings of the National Academy of Sciences of the United States of America, 112*(21), 6730–6735.

Albright, T. D. (1989). Centrifugal directional bias in the middle temporal visual area (MT) of the macaque. *Visual Neuroscience, 2*(2), 177–188.

Anzai, A., Peng, X., & Van Essen, D. C. (2007). Neurons in monkey visual area V2 encode combinations of orientations. *Nature Neuroscience, 10*(10), 1313–1321.

Arcaro, M. J., & Livingstone, M. S. (2017). A hierarchical, retinotopic proto-organization of the primate visual system at birth. *Elife, 6*.

Arcaro, M. J., Schade, P. F., Vincent, J. L., Ponce, C. R., & Livingstone, M. S. (2017). Seeing faces is necessary for face-domain formation. *Nature Neuroscience, 20*, 1404.

Arroyo, D. A., & Feller, M. B. (2016). Spatiotemporal features of retinal waves instruct the wiring of the visual circuitry. *Frontiers in Neural Circuits, 10*, 54.

Baden, T., Berens, P., Franke, K., Roman Roson, M., Bethge, M., & Euler, T. (2016). The functional diversity of retinal ganglion cells in the mouse. *Nature, 529*(7586), 345–350.

Ball, K., & Sekuler, R. (1982). A specific and enduring improvement in visual motion discrimination. *Science, 218*(4573), 697–698.

Ball, K., & Sekuler, R. (1987). Direction-specific improvement in motion discrimination. *Vision Research, 27*(6), 953–965.

Barone, I., Novelli, E., & Strettoi, E. (2014). Long-term preservation of cone photoreceptors and visual acuity in rd10 mutant mice exposed to continuous environmental enrichment. *Molecular Vision, 20,* 1545–1556.

Behrens, C., Schubert, T., Haverkamp, S., Euler, T., & Berens, P. (2016). Connectivity map of bipolar cells and photoreceptors in the mouse retina. *Elife, 5.*

Bell, A. J., & Sejnowski, T. J. (1997). The "independent components" of natural scenes are edge filters. *Vision Research, 37*(23), 3327–3338.

Bengio, Y. (2016). Machines who learn. *Scientific American, 314*(6), 46–51.

Bengio, Y., Courville, A., & Vincent, P. (2013). Representation learning: A review and new perspectives. *IEEE Transactions on Pattern Analysis and Machine Intelligence, 35*(8), 1798–1828.

Berry, K. P., & Nedivi, E. (2016). Experience-dependent structural plasticity in the visual system. *Annual Review of Vision Science, 2,* 17–35.

Besharse, J., & Bok, D. (Eds.). (2011). *The retina and its disorders.* San Diego, CA: Academic Press.

Blakemore, C., & Van Sluyters, R. C. (1975). Innate and environmental factors in the development of the kitten's visual cortex. *Journal of Physiology, 248*(3), 663–716.

Bliss, T. V., & Lomo, T. (1973). Long-lasting potentiation of synaptic transmission in the dentate area of the anaesthetized rabbit following stimulation of the perforant path. *Journal of Physiology, 232*(2), 331–356.

Bojarski, M., Del Testa, D., Dworakowski, D., Firner, B., Flepp, B., Goyal, P., et al. (2016). End to end learning for self-driving cars. arXiv e-prints. Retrieved from https://ui.adsabs.harvard.edu/abs/2016ar Xiv160407316B.

Born, R. T., & Bradley, D. C. (2005). Structure and function of visual area MT. *Annual Review of Neuroscience, 28,* 157–189.

Boycott, B. B. (2001). Brian B. Boycott. In L. R. Squire (Ed.), *The history of neuroscience in autobiography*, volume 3. Cambridge, MA: Academic Press.

Boycott, B., & Wässle, H. (1999). Parallel processing in the mammalian retina: The Proctor Lecture. *Investigative Ophthalmology and Visual Science, 40*(7), 1313–1327.

Britten, K. H. (2008). Mechanisms of self-motion perception. *Annual Review of Neuroscience, 31*, 389–410.

Brown, R. E., & Milner, P. M. (2003). The legacy of Donald O. Hebb: More than the Hebb Synapse. *Nature Reviews Neuroscience, 4*, 1013.

Butts, D. A., Kanold, P. O., & Shatz, C. J. (2007). A burst-based "Hebbian" learning rule at retinogeniculate synapses links retinal waves to activity-dependent refinement. *PLoS Biology, 5*(3), e61.

Campbell, M. (2018). Mastering board games. *Science, 362*(6419), 1118.

Cang, J., Renteria, R. C., Kaneko, M., Liu, X., Copenhagen, D. R., & Stryker, M. P. (2005). Development of precise maps in visual cortex requires patterned spontaneous activity in the retina. *Neuron, 48*(5), 797–809.

Carandini, M. (2006). What simple and complex cells compute. *Journal of Physiology, 577*(Pt 2), 463–466.

Chang, L., & Tsao, D. Y. (2017). The code for facial identity in the primate brain. *Cell, 169*(6), 1013–1028 e1014.

Chapman, B., & Stryker, M. P. (1993). Development of orientation selectivity in ferret visual cortex and effects of deprivation. *Journal of Neuroscience, 13*(12), 5251–5262.

Chatterjee, R. (2015). Out of the darkness. *Science, 350*(6259), 372–375.

Chen, J., Yamahachi, H., & Gilbert, C. D. (2010). Experience-dependent gene expression in adult visual cortex. *Cerebral Cortex, 20*(3), 650–660.

Cohen, E., & Sterling, P. (1990). Demonstration of cell types among cone bipolar neurons of cat retina. *Philosophical Transactions of the Royal Society of London. Series B, Biological Sciences, 330*(1258), 305–321.

Coimbra, J. P., Marceliano, M. L., Andrade-da-Costa, B. L., & Yamada, E. S. (2006). The retina of tyrant flycatchers: Topographic organization of neuronal density and size in the ganglion cell layer of the great kiskadee *Pitangus sulphuratus* and the rusty margined flycatcher *Myiozetetes cayanensis* (Aves: Tyrannidae). *Brain, Behavior and Evolution, 68*(1), 15–25.

Costandi, M. (2009, February 10). Where are old memories stored in the brain? *Scientific American*. Retrieved from https://www.scientific american.com/article/the-memory-trace.

Crist, R. E., Kapadia, M. K., Westheimer, G., & Gilbert, C. D. (1997). Perceptual learning of spatial localization: Specificity for orientation, position, and context. *Journal of Neurophysiology, 78*(6), 2889–2894.

Dahne, S., Wilbert, N., & Wiskott, L. (2014). Slow feature analysis on retinal waves leads to V1 complex cells. *PLOS Computational Biology, 10*(5), e1003564.

Das, S. (2017). CNN architectures: LeNet, AlexNet, VGG, GoogLeNet, ResNet and more. Retrieved from https://medium.com/@sidereal /cnns-architectures-lenet-alexnet-vgg-googlenet-resnet-and-more -666091488df5.

Daw, N. (2006). *Visual development* (2nd ed.). New York: Springer.

Denk, W., Briggman, K. L., & Helmstaedter, M. (2012). Structural neurobiology: Missing link to a mechanistic understanding of neural computation. *Nature Reviews Neuroscience, 13*(5), 351–358.

DiCarlo, J. J., Zoccolan, D., & Rust, N. C. (2012). How does the brain solve visual object recognition? *Neuron, 73*(3), 415–434.

Dolan, T., & Fernandez-Juricic, E. (2010). Retinal ganglion cell topography of five species of ground-foraging birds. *Brain, Behavior and Evolution, 75*(2), 111–121.

Dormal, G., Lepore, F., & Collignon, O. (2012). Plasticity of the dorsal "spatial" stream in visually deprived individuals. *Neural Plasticity, 2012*, 659–687.

Dowling, J. E. (2012). *The retina: An approachable part of the brain*. Cambridge, MA: Harvard University Press.

Dowling, J. E., & Dowling, J. L. (2016). *Vision: How it works and what can go wrong*. Cambridge, MA: MIT Press.

Driscoll, L. N., Pettit, N. L., Minderer, M., Chettih, S. N., & Harvey, C. D. (2017). Dynamic reorganization of neuronal activity patterns in parietal cortex. *Cell, 170*(5), 986–999 e916.

Dvorak, D., Mark, R., & Reymond, L. (1983). Factors underlying falcon grating acuity. *Nature, 303*(5919), 729–730.

Eickhoff, S. B., Yeo, B. T. T., & Genon, S. (2018). Imaging-based parcellations of the human brain. *Nature Reviews Neuroscience, 19*(11), 672–686.

Eliot, V. (Ed.) (1971). *The Waste Land: A Facsimile and Transcript of the Original Drafts*. New York: Houghton Mifflin.

El-Shamayleh, Y., Kumbhani, R. D., Dhruv, N. T., & Movshon, J. A. (2013). Visual response properties of V1 neurons projecting to V2 in macaque. *Journal of Neuroscience, 33*(42), 16594–16605.

Escher, S. A., Tucker, A. M., Lundin, T. M., & Grabiner, M. D. (1998). Smokeless tobacco, reaction time, and strength in athletes. *Medicine and Science in Sports and Exercise, 30*(10), 1548–1551.

Espinosa, J. S., & Stryker, M. P. (2012). Development and plasticity of the primary visual cortex. *Neuron, 75*(2), 230–249.

Euler, T., Detwiler, P. B., & Denk, W. (2002). Directionally selective calcium signals in dendrites of starburst amacrine cells. *Nature, 418*(6900), 845–852.

Euler, T., & Wässle, H. (1995). Immunocytochemical identification of cone bipolar cells in the rat retina. *Journal of Comparative Neurology, 361*(3), 461–478.

Fisher, C., & Freiwald, W. A. (2015). Whole-agent selectivity within the macaque face-processing system. *Proceedings of the National Academy of Sciences, 112*(47), 14717–14722.

Fite, K. V., & Rosenfield-Wessels, S. (1975). A comparative study of deep avian foveas. *Brain, Behavior and Evolution, 12*(1–2), 97–115.

Fox, R., Lehmkuhle, S. W., & Westendorf, D. H. (1976). Falcon visual acuity. *Science, 192*(4236), 263–265.

Freeman, J., Field, G. D., Li, P. H., Greschner, M., Gunning, D. E., Mathieson, K., et al. (2015). Mapping nonlinear receptive field structure in primate retina at single cone resolution. *Elife, 4.*

Freeman, J., & Simoncelli, E. P. (2011). Metamers of the ventral stream. *Nature Neuroscience, 14*(9), 1195–1201.

Freiwald, W. A., & Tsao, D. Y. (2010). Functional compartmentalization and viewpoint generalization within the macaque face-processing system. *Science, 330*(6005), 845–851.

Gaffney, M. F., & Hodos, W. (2003). The visual acuity and refractive state of the American kestrel (*Falco sparverius*). *Vision Research, 43*(19), 2053–2059.

Gandhi, T. K., Ganesh, S., & Sinha, P. (2014). Improvement in spatial imagery following sight onset late in childhood. *Psychological Science, 25*(3), 693–701.

Gandhi, T. K., Singh, A. K., Swami, P., Ganesh, S., & Sinha, P. (2017). Emergence of categorical face perception after extended early-onset blindness. *Proceedings of the National Academy of Sciences of the United States of America, 114*(23), 6139–6143.

Gattass, R., Lima, B., Soares, J. G., & Ungerleider, L. G. (2015). Controversies about the visual areas located at the anterior border of area V2 in primates. *Visual Neuroscience, 32,* E019.

Gauthier, J. L., Field, G. D., Sher, A., Greschner, M., Shlens, J., Litke, A. M., & Chichilnisky, E. J. (2009). Receptive fields in primate retina are coordinated to sample visual space more uniformly. *PLoS Biology, 7*(4), e1000063.

Gauthier, I., & Tarr, M. J. (2016). Visual object recognition: Do we (finally) know more now than we did? *Annual Review of Vision Science, 2,* 377–396.

Gegenfurtner, K. R., Kiper, D. C., & Levitt, J. B. (1997). Functional properties of neurons in macaque area V3. *Journal of Neurophysiology, 77*(4), 1906–1923.

Geitgey, A. (2016, July 24). Machine learning is fun! Part 4: Modern face recognition with deep learning. *Medium Artificial Intelligence.* Retrieved from https://medium.com/@ageitgey/machine-learning-is-fun-part-4 -modern-face-recognition-with-deep-learning-c3cffc121d78.

Ghim, M. M., & Hodos, W. (2006). Spatial contrast sensitivity of birds. *Journal of Comparative Physiology. A, Neuroethology, Sensory, Neural, and Behavioral Physiology, 192*(5), 523–534.

Ghose, G. M., Yang, T., & Maunsell, J. H. (2002). Physiological correlates of perceptual learning in monkey V1 and V2. *Journal of Neurophysiology, 87*(4), 1867–1888.

Gilbert, C. D., & Li, W. (2012). Adult visual cortical plasticity. *Neuron, 75*(2), 250–264.

Gollisch, T., & Meister, M. (2010). Eye smarter than scientists believed: Neural computations in circuits of the retina. *Neuron, 65*(2), 150–164.

Gopnik, A. (2017). Making AI more human. *Scientific American, 316*(6), 60–65.

Gopnik, A. (2019, February 22). Will A.I. ever be smarter than a four-year-old? Smithsonian.com. Retrieved from https://www.smithsonian mag.com/innovation/will-ai-ever-be-smarter-than-four-year-old -180971259.

Grady, C. L., Mondloch, C. J., Lewis, T. L., & Maurer, D. (2014). Early visual deprivation from congenital cataracts disrupts activity and functional connectivity in the face network. *Neuropsychologia, 57,* 122–139.

Gregory, R. L. (1997). *Eye and brain: The psychology of seeing* (5th ed.). Princeton, NJ: Princeton University Press.

Grens, K. (2014, November 1). A face to remember. *The Scientist*. Retrieved from https://www.the-scientist.com/cover-story/a-face-to-remember-36508.

Grimaldi, P., Saleem, K. S., & Tsao, D. (2016). Anatomical connections of the functionally defined "face patches" in the macaque monkey. *Neuron, 90*(6), 1325–1342.

Grimes, W. N., Songco-Aguas, A., & Rieke, F. (2018). Parallel processing of rod and cone signals: Retinal function and human perception. *Annual Review of Vision Science, 4*, 123–141.

Guillery, R. W. (2014). The lateral geniculate nucleus and pulvinar. In Werner, J. S., & Chalupa, L. M. (Eds.), *The new visual neurosciences* (pp. 257–283). Cambridge, MA: MIT Press.

Güntürkün, O. (1999). Sensory physiology: Vision. In G. Whittow (Ed.), *Sturkie's avian physiology* (5th ed., pp. 1–19). Cambridge MA: Academic Press.

Hammond, P. (1974). Cat retinal ganglion cells: Size and shape of receptive field centres. *Journal of Physiology, 242*(1), 99–118.

Hebb, D. O. (1949). *The organization of behavior: A neuropsychological theory*. New York: Wiley.

Hebb, D. O. (1980). D. O. Hebb. In G. Lindzey (Ed.), *A history of psychology in autobiography*, vol. VII (pp. 273–303). San Francisco: W. H. Freeman.

Helmstaedter, M., Briggman, K. L., Turaga, S. C., Jain, V., Seung, H. S., & Denk, W. (2013). Connectomic reconstruction of the inner plexiform layer in the mouse retina. *Nature, 500*(7461), 168–174.

Hinton, G., Deng, L., Yu, D., Dahl, G. E., Mohamed, A., Jaitly, N., et al. (2012). Deep neural networks for acoustic modeling in speech recognition: The shared views of four research groups. *IEEE Signal Processing Magazine, 29*(6), 82–97.

Hodos, W., Ghim, M. M., Potocki, A., Fields, J. N., & Storm, T. (2002). Contrast sensitivity in pigeons: A comparison of behavioral and pattern ERG methods. *Documenta Ophthalmologica, 104*(1), 107–118.

Holcombe, A. O. (2010). Binding problem. In E. B. Goldstein (Ed.), *Encyclopedia of Perception* (pp. 206–208). Thousand Oaks, CA: SAGE Publications.

Hubel, D. (1995). *Eye, brain, and vision*. New York: Scientific American.

Huberman, A. D., Feller, M. B., & Chapman, B. (2008). Mechanisms underlying development of visual maps and receptive fields. *Annual Review of Neuroscience, 31*, 479–509.

Hung, C. P., Kreiman, G., Poggio, T., & DiCarlo, J. J. (2005). Fast read-out of object identity from macaque inferior temporal cortex. *Science, 310*(5749), 863–866.

Ings, S. (2007). *The eye: A natural history.* London: Bloomsbury.

Inzunza, O., Bravo, H., Smith, R. L., & Angel, M. (1991). Topography and morphology of retinal ganglion cells in Falconiforms: A study on predatory and carrion-eating birds. *Anatomical Record, 229*(2), 271–277.

Issa, E. B., & DiCarlo, J. J. (2012). Precedence of the eye region in neural processing of faces. *Journal of Neuroscience, 32*(47), 16666–16682.

Jacoby, J., & Schwartz, G. W. (2017). Three small-receptive-field ganglion cells in the mouse retina are distinctly tuned to size, speed, and object motion. *Journal of Neuroscience, 37*(3), 610–625.

Kalloniatis, M., & Luu, C. (2007). Visual acuity. *Webvision.* Retrieved from https://webvision.med.utah.edu/book/part-viii-psychophysics-of-vision/visual-acuity.

Kandel, E. (2001). Nobel Lecture: The molecular biology of memory storage: A dialog between genes and synapses. *Bioscience Reports, 21,* 565–611.

Kaneko, M., & Stryker, M. P. (2017). Homeostatic plasticity mechanisms in mouse V1. *Philosophical Transactions of the Royal Society of London. Series B, Biological Sciences, 372*(1715).

Kiani, R., Esteky, H., Mirpour, K., & Tanaka, K. (2007). Object category structure in response patterns of neuronal population in monkey inferior temporal cortex. *Journal of Neurophysiology, 97*(6), 4296–4309.

Kienker, P. K., Sejnowski, T. J., Hinton, G. E., & Schumacher, L. E. (1986). Separating figure from ground with a parallel network. *Perception, 15*(2), 197–216.

Kirkby, L. A., Sack, G. S., Firl, A., & Feller, M. B. (2013). A role for correlated spontaneous activity in the assembly of neural circuits. *Neuron, 80*(5), 1129–1144.

Koch, C. (2012). *Consciousness: Confessions of a romantic reductionist.* Cambridge, MA: MIT Press.

Kolb, H. (2006). Facts and figures concerning the human retina. *Webvision.* Retrieved from https://webvision.med.utah.edu/book/part-xiii-facts-and-figures-concerning-the-human-retina.

Kornblith, S., & Tsao, D. Y. (2017). How thoughts arise from sights: Inferotemporal and prefrontal contributions to vision. *Current Opinion in Neurobiology, 46,* 208–218.

Krauzlis, R. J., Lovejoy, L. P., & Zenon, A. (2013). Superior colliculus and visual spatial attention. *Annual Review of Neuroscience, 36,* 165–182.

Krieger, B., Qiao, M., Rousso, D. L., Sanes, J. R., & Meister, M. (2017). Four alpha ganglion cell types in mouse retina: Function, structure, and molecular signatures. *PLoS One, 12*(7), e0180091.

Kumano, H., & Uka, T. (2013). Neuronal mechanisms of visual perceptual learning. *Behavioural Brain Research, 249,* 75–80.

Lashley, K. S. (1950). In search of the engram. *Physiological mechanisms in animal behavior (Society's Symposium IV)* (pp. 454–482). Oxford, UK: Academic Press.

LeCun, Y., Bengio, Y., & Hinton, G. (2015). Deep learning. *Nature, 521*(7553), 436–444.

Lewis-Kraus, G. (2016, December 14). The great A.I. awakening. *New York Times Magazine.* Retrieved from https://www.nytimes.com /2016/12/14/magazine/the-great-ai-awakening.html.

Li, S. Z., & Jain, A. (Eds.). (2011). *Handbook of face recognition* (2nd ed.). New York: Springer.

Lindsey, J., Ocko, S. A., Ganguli, S., & Deny, S. (2019). A unified theory of early visual representations from retina to cortex through anatomically constrained deep CNNs. arXiv e-prints. Retrieved from https://ui.adsabs.harvard.edu/abs/2019arXiv190100945L.

Litvina, E. Y., & Chen, C. (2017). Functional convergence at the retinogeniculate synapse. *Neuron, 96*(2), 330–338 e335.

Liu, L., She, L., Chen, M., Liu, T., Lu, H. D., Dan, Y., & Poo, M. M. (2016). Spatial structure of neuronal receptive field in awake monkey secondary visual cortex (V2). *Proceedings of the National Academy of Sciences of the United States of America, 113*(7), 1913–1918.

Liu, Y. S., Stevens, C. F., & Sharpee, T. O. (2009). Predictable irregularities in retinal receptive fields. *Proceedings of the National Academy of Sciences of the United States of America, 106*(38), 16499–16504.

Livingstone, M. S., Pack, C. C., & Born, R. T. (2001). Two-dimensional substructure of MT receptive fields. *Neuron, 30*(3), 781–793.

Livingstone, M. S., Vincent, J. L., Arcaro, M. J., Srihasam, K., Schade, P. F., & Savage, T. (2017). Development of the macaque face-patch system. *Nature Communications, 8,* 14897.

MacNeil, M. A., Heussy, J. K., Dacheux, R. F., Raviola, E., & Masland, R. H. (1999). The shapes and numbers of amacrine cells: Matching of photofilled with Golgi-stained cells in the rabbit retina and comparison with other mammalian species. *Journal of Comparative Neurology, 413,* 305–326.

MacNeil, M. A., Heussy, J. K., Dacheux, R. F., Raviola, E., & Masland, R. H. (2004). The population of bipolar cells in the rabbit retina. *Journal of Comparative Neurology, 472,* 73–86.

Margolis, D. J., Lutcke, H., Schulz, K., Haiss, F., Weber, B., Kugler, S., et al. (2012). Reorganization of cortical population activity imaged throughout long-term sensory deprivation. *Nature Neuroscience, 15*(11), 1539–1546.

Martersteck, E. M., Hirokawa, K. E., Evarts, M., Bernard, A., Duan, X., Li, Y., et al. (2017). Diverse central projection patterns of retinal ganglion cells. *Cell Reports, 18*(8), 2058–2072.

Martin, P., & Masland, R. H. (2007). Essay: The unsolved mystery of vision. *Current Biology* 18:R577–R583.

Masland, R. H. (2001). Neuronal diversity in the retina. *Current Opinion in Neurobiology, 11,* 431–436.

Masland, R. H. (2012). The neuronal organization of the retina. *Neuron, 76,* 266–280.

McGuire, B. A., Stevens, J. K., & Sterling, P. (1984). Microcircuitry of bipolar cells in cat retina. *Journal of Neuroscience, 4*(12), 2920–2938.

McKyton, A., Ben-Zion, I., Doron, R., & Zohary, E. (2015). The limits of shape recognition following late emergence from blindness. *Current Biology, 25*(18), 2373–2378.

McMahan, U. (1990). *Steve: Remembrances of Stephen W. Kuffler.* Sunderland, MA: Sinauer Associates.

McManus, J. N., Li, W., & Gilbert, C. D. (2011). Adaptive shape processing in primary visual cortex. *Proceedings of the National Academy of Sciences of the United States of America, 108*(24), 9739–9746.

Meyers, E. M., Borzello, M., Freiwald, W. A., & Tsao, D. (2015). Intelligent information loss: The coding of facial identity, head pose, and non-face information in the macaque face patch system. *Journal of Neuroscience, 35*(18), 7069–7081.

Moeller, S., Crapse, T., Chang, L., & Tsao, D. Y. (2017). The effect of face patch microstimulation on perception of faces and objects. *Nature Neuroscience, 20*(5), 743–752.

Montañez, A. (2016, May 20). Unveiling the hidden layers of deep learning. *SA Visual,* a blog in *Scientific American.* Retrieved from https://blogs.scientificamerican.com/sa-visual/unveiling-the-hidden-layers-of-deep-learning.

Moore, B. A., Tyrrell, L. P., Pita, D., Bininda-Emonds, O. R. P., & Fernández-Juricic, E. (2017). Does retinal configuration make the head and eyes of foveate birds move? *Scientific Reports, 7,* 38406.

Moore, B. D., Kiley, C. W., Sun, C., & Usrey, W. M. (2011). Rapid plasticity of visual responses in the adult lateral geniculate nucleus. *Neuron, 71*(5), 812–819.

Morgan, J. L., Berger, D. R., Wetzel, A. W., & Lichtman, J. W. (2016). The fuzzy logic of network connectivity in mouse visual thalamus. *Cell, 165*(1), 192–206.

Movshon, J. A., Lisberger, S. G., & Krauzlis, R. J. (1990). Visual cortical signals supporting smooth pursuit eye movements. *Cold Spring Harbor Symposia on Quantitative Biology, 55*, 707–716.

Movshon, J. A., & Newsome, W. T. (1996). Visual response properties of striate cortical neurons projecting to area MT in macaque monkeys. *Journal of Neuroscience, 16*(23), 7733–7741.

Movshon, J. A., & Simoncelli, E. P. (2014). Representation of naturalistic image structure in the primate visual cortex. *Cold Spring Harbor Symposia on Quantitative Biology, 79*, 115–122.

Ohki, K., Chung, S., Ch'ng, Y. H., Kara, P., & Reid, R. C. (2005). Functional imaging with cellular resolution reveals precise microarchitecture in visual cortex. *Nature, 433*(7026), 597–603.

O'Keefe, J. (2014, December 7). Spatial cells in the hippocampal formation. Nobel Lecture.

O'Keefe, J., & Dostrovsky, J. (1971). The hippocampus as a spatial map: Preliminary evidence from unit activity in the freely-moving rat. *Brain Research, 34*(1), 171–175.

Olshausen, B. A., & Field, D. J. (1996). Natural image statistics and efficient coding. *Network, 7*(2), 333–339.

O'Rourke, C. T., Hall, M. I., Pitlik, T., & Fernandez-Juricic, E. (2010). Hawk eyes I: Diurnal raptors differ in visual fields and degree of eye movement. *PLoS One, 5*(9), e12802.

Pack, C. C., & Born, R. T. (2004). Responses of MT neurons to barber pole stimuli. *Journal of Vision, 4*(859).

Pack, C. C., & Born, R. (2010). Cortical mechanisms for the integration of visual motion. In R. H. Masland, T. D. Albright, G. M. Shephard, & E. P. Gardner (Eds.), *The senses: A comprehensive reference*, vol. 2 (pp. 189–218). San Diego, CA: Academic Press.

Pack, C. C., Gartland, A. J., & Born, R. T. (2004). Integration of contour and terminator signals in visual area MT of alert macaque. *Journal of Neuroscience, 24*(13), 3268–3280.

Peron, S. P., Freeman, J., Iyer, V., Guo, C., & Svoboda, K. (2015). A cellular resolution map of barrel cortex activity during tactile behavior. *Neuron, 86*(3), 783–799.

Phillips, P. J., Grother, P., Michaels, R. J., Balackburn, D. M., Tabassi, E., & Bone, M. (2003). *Face recognition vendor test 2002: Evaluation report* (6965). Retrieved from https://nvlpubs.nist.gov/nistpubs/Legacy/IR /nistir6965.pdf.

Ponce, C. R., Hartmann, T. S., & Livingstone, M. S. (2017). End-stopping predicts curvature tuning along the ventral stream. *Journal of Neuroscience, 37*(3), 648–659.

Pritchard, R. M., Heron, W., & Hebb, D. O. (1960). Visual perception approached by the method of stabilized images. *Canadian Journal of Experimental Psychology, 14*, 67–77.

Protti, D. A., Flores-Herr, N., Li, W., Massey, S. C., & Wässle, H. (2005). Light signaling in scotopic conditions in the rabbit, mouse and rat retina: A physiological and anatomical study. *Journal of Neurophysiology, 93*(6), 3479–3488.

Quiroga, R. Q., Reddy, L., Kreiman, G., Koch, C., & Fried, I. (2005). Invariant visual representation by single neurons in the human brain. *Nature, 435*(7045), 1102–1107.

Raiguel, S., Van Hulle, M. M., Xiao, D. K., Marcar, V. L., & Orban, G. A. (1995). Shape and spatial distribution of receptive fields and antagonistic motion surrounds in the middle temporal area (V5) of the macaque. *European Journal of Neuroscience, 7*(10), 2064–2082.

Reid, R. C., & Alonso, J. M. (1995). Specificity of monosynaptic connections from thalamus to visual cortex. *Nature, 378*(6554), 281–284.

Reid, R. C., & Usrey, W. M. (2013). Vision. In L. Squire, D. Berg, F. E. Bloom, S. du Lac, A. Ghosh, & N. C. Spitzer (Eds.), *Fundamental neuroscience* (4th ed., pp. 577–595). Oxford, UK: Academic Press.

Reymond, L. (1985). Spatial visual acuity of the eagle *Aquila audax*: A behavioural, optical and anatomical investigation. *Vision Research, 25*(10), 1477–1491.

Reymond, L. (1987). Spatial visual acuity of the falcon, *Falco berigora*: A behavioural, optical and anatomical investigation. *Vision Research, 27*(10), 1859–1874.

Richert, M., Albright, T. D., & Krekelberg, B. (2013). The complex structure of receptive fields in the middle temporal area. *Frontiers in Systems Neuroscience, 7*, 2.

Riesenhuber, M., & Poggio, T. (1999). Hierarchical models of object recognition in cortex. *Nature Neuroscience, 2*(11), 1019–1025.

Roe, A. W., Chelazzi, L., Connor, C. E., Conway, B. R., Fujita, I., Gallant, J. L., et al. (2012). Toward a unified theory of visual area V4. *Neuron, 74*(1), 12–29.

Roelfsema, P. R., & Holtmaat, A. (2018). Control of synaptic plasticity in deep cortical networks. *Nature Reviews Neuroscience, 19*(3), 166–180.

Roman Roson, M., Bauer, Y., Kotkat, A. H., Berens, P., Euler, T., & Busse, L. (2019). Mouse dLGN receives functional input from a diverse population of retinal ganglion cells with limited convergence. *Neuron, 102*(2), 462–476 e468.

Rompani, S. B., Mullner, F. E., Wanner, A., Zhang, C., Roth, C. N., Yonehara, K., & Roska, B. (2017). Different modes of visual integration in the lateral geniculate nucleus revealed by single-cell-initiated transsynaptic tracing. *Neuron, 93*(4), 767–777.

Rose, T., & Bonhoeffer, T. (2018). Experience-dependent plasticity in the lateral geniculate nucleus. *Current Opinion in Neurobiology, 53,* 22–28.

Roska, B. (2019). The first steps in vision: Cell types, circuits, and repair. *EMBO Molecular Medicine, 11*(3).

Rossion, B., & Taubert, J. (2017). Commentary: The code for facial identity in the primate brain. *Frontiers in Human Neuroscience, 11,* 550.

Ruggeri, M., Major, J. C. Jr., McKeown, C., Knighton, R. W., Puliafito, C. A., & Jiao, S. (2010). Retinal structure of birds of prey revealed by ultra-high resolution spectral-domain optical coherence tomography. *Investigative Ophthalmology and Visual Science, 51*(11), 5789–5795.

Sagi, D. (2011). Perceptual learning in vision research. *Vision Research, 51*(13), 1552–1566.

Sanes, J. R., & Masland, R. H. (2015). The types of retinal ganglion cells: Current status and implications for neuronal classification. *Annual Review of Neuroscience, 38,* 221–246.

Scholl, B., & Priebe, N. J. (2015). Neuroscience: The cortical connection. *Nature, 518*(7539), 306–307.

Seabrook, T. A., Burbridge, T. J., Crair, M. C., & Huberman, A. D. (2017). Architecture, function, and assembly of the mouse visual system. *Annual Review of Neuroscience, 40,* 499–538.

Sejnowski, T. J. (2018). *The deep learning revolution: Artificial intelligence meets human intelligence.* Cambridge, MA: MIT Press.

Seung, S. (2012). *Connectome: How the brain's wiring makes us who we are.* New York: Houghton Mifflin Harcourt.

Shadlen, M. N., & Movshon, J. A. (1999). Synchrony unbound: A critical evaluation of the temporal binding hypothesis. *Neuron, 24*(1), 67–77, 111–125.

Sharma, J., Angelucci, A., & Sur, M. (2000). Induction of visual orientation modules in auditory cortex. *Nature, 404*(6780), 841–847.

Sheikh, K. (2017, June 1). How we save face—Researchers crack the brain's facial-recognition code. *Scientific American.* Retrieved from https://www.scientificamerican.com/article/how-we-save-face-researchers-crack-the-brains-facial-recognition-code.

Shekhar, K., Lapan, S. W., Whitney, I. E., Tran, N. M., Macosko, E. Z., Kowalczyk, M., et al. (2016). Comprehensive classification of retinal bipolar neurons by single-cell transcriptomics. *Cell, 166*(5), 1308–1323 e1330.

Sherman, S. M., & Guillery, R. W. (2013). *Functional connections of cortical areas.* Cambridge, MA: MIT Press.

Silver, D., Hubert, T., Schrittwieser, J., Antonoglou, I., Lai, M., Guez, A., et al. (2018). A general reinforcement learning algorithm that masters chess, shogi, and Go through self-play. *Science, 362*(6419), 1140–1144.

Sincich, L. C., Horton, J. C., & Sharpee, T. O. (2009). Preserving information in neural transmission. *Journal of Neuroscience, 29*(19), 6207–6216.

Sinha, P. (2013). Once blind and now they see. *Scientific American, 309*(1), 48–55.

Smolyanskaya, A., Haefner, R. M., Lomber, S. G., & Born, R. T. (2015). A modality-specific feedforward component of choice-related activity in MT. *Neuron, 87*(1), 208–219.

Solomon, S. G., Tailby, C., Cheong, S. K., & Camp, A. J. (2010). Linear and nonlinear contributions to the visual sensitivity of neurons in primate lateral geniculate nucleus. *Journal of Neurophysiology, 104*(4), 1884–1898.

Srihasam, K., Vincent, J. L., & Livingstone, M. S. (2014). Novel domain formation reveals proto-architecture in inferotemporal cortex. *Nature Neuroscience, 17*(12), 1776–1783.

Stevens, C. F. (1998). Neuronal diversity: Too many cell types for comfort? *Current Biology, 8*(20), R708–710.

Stokkan, K. A., Folkow, L., Dukes, J., Neveu, M., Hogg, C., Siefken, S., et al. (2013). Shifting mirrors: Adaptive changes in retinal reflections to winter darkness in Arctic reindeer. *Proceedings of the Royal Society B: Biological Sciences, 280*(1773), 20132451.

Strogatz, S. (2018, December 26). One giant step for a chess-playing machine. *New York Times.* Retrieved from https://www.nytimes.com/2018/12/26/science/chess-artificial-intelligence.html.

Strom, R. C. (1999). Genetic and environmental control of retinal ganglion cell variation. Chapter 4 in *Genetic analysis of variation in neuron number*, PhD diss., University of Tennessee Health Science Center, Memphis, Tennessee. Retrieved from www.nervenet.org/papers /strom99/Chapter4.html.

Sumbul, U., Song, S., McCulloch, K., Becker, M., Lin, B., Sanes, J. R., et al. (2014). A genetic and computational approach to structurally classify neuronal types. *Nature Communications, 5*, 3512.

Suresh, V., Ciftcioglu, U. M., Wang, X., Lala, B. M., Ding, K. R., Smith, W. A., et al. (2016). Synaptic contributions to receptive field structure and response properties in the rodent lateral geniculate nucleus of the thalamus. *Journal of Neuroscience, 36*(43), 10949–10963.

Tanaka, K. (1983). Cross-correlation analysis of geniculostriate neuronal relationships in cats. *Journal of Neurophysiology, 49*(6), 1303–1318.

Tanaka, K. (1985). Organization of geniculate inputs to visual cortical cells in the cat. *Vision Research, 25*(3), 357–364.

Tang, S., Lee, T. S., Li, M., Zhang, Y., Xu, Y., Liu, F., et al. (2018). Complex pattern selectivity in macaque primary visual cortex revealed by large-scale two-photon imaging. *Current Biology, 28*(1), 38–48 e33.

Thompson, A., Gribizis, A., Chen, C., & Crair, M. C. (2017). Activity-dependent development of visual receptive fields. *Current Opinion in Neurobiology, 42*, 136–143.

Tien, N. W., Pearson, J. T., Heller, C. R., Demas, J., & Kerschensteiner, D. (2015). Genetically identified suppressed-by-contrast retinal ganglion cells reliably signal self-generated visual stimuli. *Journal of Neuroscience, 35*(30), 10815–10820.

Tonegawa, S., Liu, X., Ramirez, S., & Redondo, R. (2015). Memory engram cells have come of age. *Neuron, 87*(5), 918–931.

Tootell, R. B., Reppas, J. B., Dale, A. M., Look, R. B., Sereno, M. I., Malach, R., et al. (1995). Visual motion aftereffect in human cortical area MT revealed by functional magnetic resonance imaging. *Nature, 375*(6527), 139–141.

Tsao, D. (2014). The macaque face patch system: A window into object representation. *Cold Spring Harbor Symposia on Quantitative Biology, 79*, 109–114.

Tsao, D. Y., & Livingstone, M. S. (2008). Mechanisms of face perception. *Annual Review of Neuroscience, 31*, 411–437.

Tsodyks, M., & Gilbert, C. (2004). Neural networks and perceptual learning. *Nature, 431*(7010), 775–781.

Turner, M. H., Sanchez Giraldo, L. G., Schwartz, O., & Rieke, F. (2019). Stimulus- and goal-oriented frameworks for understanding natural vision. *Nature Neuroscience, 22*(1), 15–24.

Wagner, I. C. (2016). The integration of distributed memory traces. *Journal of Neuroscience, 36*(42), 10723–10725.

Wandell, B. A., & Smirnakis, S. M. (2009). Plasticity and stability of visual field maps in adult primary visual cortex. *Nature Reviews Neuroscience, 10*(12), 873–884.

Wang, H. X., & Movshon, J. A. (2016). Properties of pattern and component direction-selective cells in area MT of the macaque. *Journal of Neurophysiology, 115*(6), 2705–2720.

Wässle, H. (2002). Brian Blundell Boycott, 10 December 1924–22 April 2000. *Biographical Memoirs of Fellows of the Royal Society, 48*, 53–68.

Wässle, H., Grunert, U., Rohrenbeck, J., & Boycott, B. B. (1989). Cortical magnification factor and the ganglion cell density of the primate retina. *Nature, 341*(6243), 643–646.

Wässle, H., Puller, C., Muller, F., & Haverkamp, S. (2009). Cone contacts, mosaics, and territories of bipolar cells in the mouse retina. *Journal of Neuroscience, 29*(1), 106–117.

Watanabe, T., Náñez, J. E., & Sasaki, Y. (2001). Perceptual learning without perception. *Nature, 413*(6858), 844–848.

Wathey, J. C., & Pettigrew, J. D. (1989). Quantitative analysis of the retinal ganglion cell layer and optic nerve of the barn owl *Tyto alba*. *Brain, Behavior and Evolution, 33*(5), 279–292.

Werner, J. S., & Chalupa, L. M. (Eds.). (2014). *The new visual neurosciences*. Cambridge, MA: The MIT Press.

Wiesel, T. N. (1982). Postnatal development of the visual cortex and the influence of environment. *Nature, 299*(5884), 583–591.

Wong, R. O., Meister, M., & Shatz, C. J. (1993). Transient period of correlated bursting activity during development of the mammalian retina. *Neuron, 11*(5), 923–938.

Wu, K. J. (2018, December 10). Google's new A.I. is a master of games, but how does it compare to the human mind? Smithsonian.com. Retrieved from www.smithsonianmag.com/innovation/google-ai-deepminds-alphazero-games-chess-and-go-180970981.

Yamins, D. L., & DiCarlo, J. J. (2016). Using goal-driven deep learning models to understand sensory cortex. *Nature Neuroscience, 19*(3), 356–365.

Yamins, D. L. K., Hong, H., Cadieu, C. F., Solomon, E. A., Seibert, D., & DiCarlo, J. J. (2014). Performance-optimized hierarchical models predict neural responses in higher visual cortex. *Proceedings of the National Academy of Sciences, 111*(23), 8619–8624.

Zeng, H., & Sanes, J. R. (2017). Neuronal cell-type classification: Challenges, opportunities and the path forward. *Nature Reviews Neuroscience, 18*(9), 530–546.

Zhang, X., Zhao, J., & LeCun, Y. (2015). Character-level convolutional networks for text classification. In Cortes, C., Lawrence, N. D., Lee, D. D., Sugiyama, M., & Garnett, R. (Eds.), *Advances in neural information processing systems 28*. Red Hook, NY: Curran.

Zimmerman, A., Bai, L., & Ginty, D. D. (2014). The gentle touch receptors of mammalian skin. *Science, 346*(6212), 950–954.

Zuccolo, R. (2017, April 3). Self-driving cars—Advanced computer vision with opencv, finding lane lines. Retrieved from https://chatbotslife
.com/self-driving-cars-advanced-computer-vision-with-opencv
-finding-lane-lines-488a411b2c3d.

Index

Richard Masland (1942–2019) was the David Glendenning Cogan Distinguished Professor of Ophthalmology and Professor of Neuroscience at Harvard Medical School. He was vice chair for research in ophthalmology at Harvard's Massachusetts Eye and Ear Infirmary, the world's largest vision research institute. For more than twenty years he was head of a major Harvard Medical School course in Neurobiology, for which he won two teaching prizes. He was a fellow of the AAAS, a former Howard Hughes Medical Institute investigator, and a recipient of the Proctor Medal and Alcon Research awards, among others. Masland made groundbreaking contributions to the study of the neural networks of the retina.